Electrodynamics

Chicago Lectures in Physics
Robert M. Wald, series editor
    Henry J. Frisch
    Gene R. Mazenko
    Sidney R. Nagel

Other *Chicago Lectures in Physics* titles available
from the University of Chicago Press

Currents and Mesons, by J. J. Sakurai (1969)

Mathematical Physics, by Robert Geroch (1984)

Useful Optics, by Walter T. Welford (1991)

*Quantum Field Theory in Curved Spacetime and
    Black Hole Thermodynamics*, by Robert M. Wald (1994)

Geometrical Vectors, by Gabriel Weinreich (1998)

Electrodynamics

Fulvio Melia

The University of Chicago Press

Chicago and London

The University of Chicago Press, Chicago 60637
The University of Chicago Press, Ltd., London
© 2001 by The University of Chicago
All rights reserved. Published 2001
Printed in the United States of America

18 17 16 15 14 13 12 11 10 09   2 3 4 5 6

ISBN-13: 978-0-226-51957-9 (cloth)
ISBN-13: 978-0-226-51958-6 (paper)
ISBN-10: 0-226-51957-0 (cloth)
ISBN-10: 0-226-51958-9 (paper)

LIBRARY OF CONGRESS CATALOGING-IN-PUBLICATION DATA

Melia, Fulvio.
    Electrodynamics / Fulvio Melia.
        p.   cm. — (Chicago lectures in physics)
    Includes bibliographical references and index.
    ISBN 0-226-51957-0 (cloth : alk. paper) — ISBN 0-226-51958-9 (pbk : alk.
    paper)
    1. Electrodynamics.   I. Title.   II. Series.

    QC631 .M445 2001
    537.6—dc21   2001023059

CONTENTS

PREFACE

This book is based on a graduate course that I have had the pleasure of teaching to six different entering classes of physics graduate students. The two-semester sequence on electrodynamics, encompassing approximately 50 lectures, was designed to lead them from the basic physical principles of this subject—e.g., the concept of a field and the *physical* meaning of a divergence and a curl—through to a relativistic Lagrangian formalism with action principles that overlap with the field theoretic techniques used in other branches of advanced physics. The course material is self-contained, requiring little (or no) reference to other texts, though several key sources cited in the book ought to be consulted for a deeper understanding of some of the topics. The main theme throughout has been to produce a concise, compact, and yet complete development of this important branch of physics, making the fundamental ideas and principles easily accessible and always strongly motivated.

Electrodynamics is a long-established discipline in physics, and several good texts are available, among them those by Jackson and by Landau and Lifshitz. However, my students have often complained that it is difficult to find a coherent and *continuous* treatment of the various subtopics. Often, a text will either begin at a basic level and end before getting to the advanced treatment of the subject (which is essential, e.g., for a coupling of the ideas in electrodynamics to the more advanced aspects of particle physics), or will begin by assuming that the reader is familiar with the basic concepts and develop the formalism from an advanced standpoint. Each of these approaches has its merit. The former permits an in-depth study of electrodynamics in the context of nonrelativistic particle dynamics, with many applications and nuances, though often losing sight of the beautiful and elegant field theoretic treatment that lies beyond. The latter exposes students to the material they will need in order to advance to a quantization of the classical field, but in starting beyond the basics, there is an inherent danger of losing the physics and experimental foundation for the theory. For example, no graduate student should have completed a course in electrodynamics without fully appreciating the physical basis for gauge invariance, or the fact that the constancy of the electron mass is an indication that a charge radiates when its distorted self-field attempts to readjust to its preaccelerated state. For the young graduate student, a self-consistent, complete transition from the empirical basis of Maxwell's equations to topics such as these is a must.

Wherever possible, I have attempted to emphasize the physics of the principles under discussion with a view toward always bridging previous and new ideas, and the mathematical formalism, in a clearly motivated manner. I have found that a sustained, motivational approach using well-defined physical concepts is the key ingredient in a successful development of the subject from its simplest steps to the full Lagrangian formalism. In my course, for example, I treat radiating systems twice—once with classical tools, and a second time with the complete, relativistic description. This serves several purposes, the most important of which is that the students gain a physical understanding of the principles of radiation using fundamental ideas and can then fully appreciate and correctly interpret what the complete, field theoretic equations are telling us.

I wish to thank Eugene N. Parker for graciously accepting the task of reading through the text. His many notes and suggestions have led to many improvements, and to him I am most grateful. I am also greatly indebted to the second (anonymous) reader, who himself made numerous suggestions for improvements and corrections that have greatly enhanced the quality of this book. Over the years, more than 150 graduate students have contributed to the refinements in my lecture notes that formed the early stages of this book. Their insightful questions, their love of physics, and their perseverance have been an inspiration to me. My appreciation for the beauty and elegance of this subject has grown with theirs over this time. I hope this monograph will do the same for others yet to come. Finally, I owe a debt of gratitude to the pillars of my life: Patricia, Marcus. Eliana. and Adrian.

1

INTRODUCTION

1.1 THE PHYSICAL BASIS OF MAXWELL'S EQUATIONS

It is often said that the theory of electrodynamics is beautiful because of its completeness and precision. It should be noted more frequently that its appeal is indeed a measure of the elegance with which such an elaborate theoretical superstructure rests firmly on a selection of physical laws that are directly derivable from a few simple observational facts. It seems appropriate, therefore, to begin our discussion by examining the physical basis of these anchoring relations—the Maxwell equations.

The subject of our study is something known as the electromagnetic field. Just one of many important fields in physics, it is a continuously defined function (or a set of functions) of the coordinates of a point in space, and sometimes, in spacetime. We shall quickly come to understand why the electromagnetic field is now well established as a dynamical entity that can interact with charges and currents, but one that can nonetheless exist on its own and carry attributes such as energy, momentum, and angular momentum. To begin with, however, let us briefly trace the first steps in our recognition of its presence, which is manifested via its coupling to particles. The first of its components, the electric field $\mathbf{E}$, is observed experimentally to be produced either by a charge Q, or by a changing magnetic field, which we define below. In a region of space where the charges are static, the electric field due to a charge Q is defined in terms of the force $\mathbf{F}_{\text{static}}$ experienced by a second (test) charge q, such that

$$\mathbf{E}(x, y, z, t) = \lim_{q \to 0} \frac{\mathbf{F}_{\text{static}}}{q} , \qquad (1.1)$$

1

where the limit simply removes any possible effect on the field due to the test charge itself. The empirically derived form of $\mathbf{F}_{static}$, i.e., Coulomb's law,

$$\mathbf{F}_{static} = \frac{qQ\hat{r}}{r^2} \, , \tag{1.2}$$

where the unit vector $\hat{r}$ points from Q toward q, was considered at the time of its discovery to be an example of action at a distance, in which two charges act on each other in a way that has nothing to do with the intervening medium. Work with dielectric polarization, pioneered by Faraday (1791-1867), showed that this is clearly not tenable, for the effect on a capacitor due to the presence of a dielectric between the two plates calls for a dependence of the electric force on the charge of the intervening medium. In other words, electric forces must be transmitted *by* the medium.

Maxwell (1831-1879) built on Faraday's ideas and succeeded in showing that the forces (and their energies) exerted on charges by other charges could be expressed not only in terms of the magnitudes of the charges themselves and their locations, but also in terms of a stress-energy tensor defined throughout the medium in terms of certain functions of the *field strengths*. This tensor—a matrix of elements that include the vector components of the momentum flux density propagating in each of the spatial dimensions—can change with location and may be used to infer the net momentum transfer across any given surface throughout the volume encompassing the particles. For example, the force between two charges could either be calculated via the empirical Coulomb's law or by integrating the stress-energy tensor over an imaginary surface surrounding the charge. With the development of Maxwell's equations the field concept was firmly in place, particularly with the subsequent discovery of electromagnetic waves and their identity with light waves.[1] We now know, of course, that the field is a *dynamical* entity; it possesses energy, momentum, and angular momentum—it is not merely a mathematical function. Indeed, the reader's ability to see this page is based entirely on the reality of photons in this field.

The basic idea of *classic* electrodynamics as a field theory is that charges and currents produce at each point of spacetime a field that has a reality of its own, and which can affect other charges. There are two sets of equations that account for these effects: Maxwell's equations describe the field produced *by* the charges, and in turn, the Lorentz force equation shows how a field acts *on* a charge. Such a simple division is not always possible. In quantum chromodynamics, for

---

[1]The reader is encouraged to learn more about the history of the field concept from the nonmathematical, but very detailed account in Williams (1966).

example, the fields themselves carry "charge" so the equations are necessarily strongly coupled and nonlinear.[2]

Let us remind ourselves of what the Lorentz force equation looks like, because it represents one method of defining the second component of the electromagnetic field (i.e., the vacuum magnetic field **B**) from empirical data:

$$\mathbf{F} = q\left(\mathbf{E} + \frac{\mathbf{v}}{c} \times \mathbf{B}\right). \tag{1.3}$$

In this case, the force experienced by the charge q has two components, one due to **E** and the second due to **B**, and it is to be distinguished from its value $\mathbf{F}_{static}$ in the static limit, which we used in Equations (1.1) and (1.2) above. Assuming we can measure **F** and that we know **E**, the field **B** is defined in terms of the force experienced by a *moving* test charge q as

$$\mathbf{v} \times \mathbf{B} = \lim_{q \to 0} \frac{c\mathbf{F}}{q} - c\mathbf{E}. \tag{1.4}$$

In cases where charges move and produce static (i.e., time-independent) currents, the magnetic field may also be identified in terms of the "source" properties by using another empirically derived equation—the Biot-Savart rule:

$$\mathbf{B}(\mathbf{x}) = \frac{Q\,\mathbf{v} \times \hat{r}}{r^2\,c}, \tag{1.5}$$

where **v** is now the (uniform) velocity of the particle producing the field. The motivation for this definition is the experimentally inferred force exerted on one current-carrying loop by a second. As is the case in electrostatics, it is convenient to imagine that one of the currents produces a field which then exerts a force on the other current. As we noted above, this definition is not imperative until we consider time-varying phenomena, but it becomes necessary then in order to preserve the conservation of energy and momentum, since the magnetic field, together with the electric field, then constitute a dynamical entity. A charge Q moving with velocity **v** is "seen" to produce a magnetic field **B** according to Equation (1.5), such that the force exerted on a second charge q is then correctly given by (the Lorentz force) Equation (1.3).

The first of Maxwell's equations, Gauss's law for the electric field, is *deduced* from Coulomb's law. It simply says that the total electric flux threading a closed surface is proportional to the net charge enclosed by that surface. As long as the fields of different charges act independently of each other, this is an intuitively obvious concept, since twice as many charges should produce twice

---

[2]See Yndurain (1983) for an introductory treatise on this subject.

as much field. The constant of proportionality is fixed by the requirement that this law correctly reproduces the empirically derived Equation (1.2):

$$\oint_S \mathbf{E} \cdot d\mathbf{a} = 4\pi \sum Q \, . \tag{1.6}$$

However, because of the complexity of most charge distributions, this integral form is not very useful. Instead, we would like to have a law that deals with the flux threading the surface enclosing an infinitesimal volume element. We are therefore motivated to introduce the *divergence* of the electric field, defined to be the *outflux per unit volume V*:

$$\operatorname{div} \mathbf{E} \equiv \lim_{V \to 0} \left\{ \frac{1}{V} \oint_{S(V)} \mathbf{E} \cdot d\mathbf{a} \right\} \, . \tag{1.7}$$

In this (and subsequent) applications of the volume limit, it is understood by the limiting symbol that V is shrunk to the point $\mathbf{x}_0$ at which the divergence is to be calculated, in such a way as to at each stage always contain $\mathbf{x}_0$ in its interior, and that $S(V)$ always denotes the boundary of V, so that both S and V change in the limit process. But what does $\operatorname{div} \mathbf{E}$ look like mathematically? Consider the case of a divergence from the cubical volume element $dx \, dy \, dz$ in a Cartesian coordinate system (Figure 1.1).

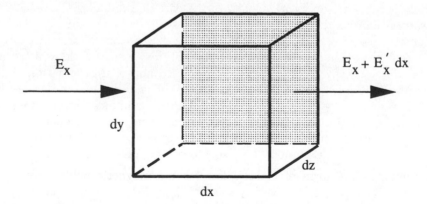

Figure 1.1   The divergence of the electric field from a cubical volume element in Cartesian coordinates. The change in E_x as the field traverses across the cube is $E'_x \, dx$, where $E'_x \equiv \partial E_x / \partial x$.

The net flux of the **E**-vector from the inside to the outside of this box is

$$\oint_S \mathbf{E} \cdot d\mathbf{a} = \sum \mathbf{E} \cdot d\mathbf{a}$$

$$= \left(\frac{\partial E_x}{\partial x} \, dx \right) dy \, dz + \left(\frac{\partial E_y}{\partial y} \, dy \right) dx \, dz + \left(\frac{\partial E_z}{\partial z} \, dz \right) dx \, dy \,, \qquad (1.8)$$

where the sum is taken over the six faces of the cube. Thus, using $V = dx \, dy \, dz$, we get

$$\text{div}\,\mathbf{E} = \frac{\partial E_x}{\partial x} + \frac{\partial E_y}{\partial y} + \frac{\partial E_z}{\partial z} \equiv \vec{\nabla} \cdot \mathbf{E} \,. \qquad (1.9)$$

Moreover, we can write

$$\sum Q = \int_V \rho \, d^3 x \,, \qquad (1.10)$$

where ρ is the charge density enclosed by the box. But

$$\lim_{V \to 0} \left\{ \frac{1}{V} \int_V \rho \, d^3 x \right\} = \rho \,, \qquad (1.11)$$

which together with Equations (1.6) and (1.7) brings us to the first of Maxwell's equations written in differential form:

$$\boxed{\text{div}\,\mathbf{E} \equiv \vec{\nabla} \cdot \mathbf{E} = 4\pi\rho} \,. \qquad (1.12)$$

Strictly speaking, our derivation of this equation is based in part on the requirement that the constant of proportionality in Equation (1.6) correctly matches that in Coulomb's law (Equation [1.2]), which describes a static field. As long as the electric fields of different charges continue to act independently of each other even in time-dependent situations, it is reasonable to assume that Equation (1.12) applies generally to any charge distribution ρ, whether or not the configuration is static.

The corresponding form of Gauss's law for the magnetic field depends on whether or not magnetic monopoles exist. Like their electric charge counterparts, monopoles are *point sources* for **B**, producing radial fields, different from other types of magnetic fields that can be represented as lines of force that do not begin or end on sources, e.g., they may be closed loops or curves that extend from minus to plus infinity. Monopoles are extremely interesting objects, whose relic abundance in the universe is unknown, and whose flux is severely constrained by several astrophysical/cosmological arguments. Because of their enormous mass, grand unified theory monopoles cannot be produced in the lab, and their detection must necessarily involve searching for cosmological particles.[3] One such search, using the induction of a current in a loop due to the passage of a magnetic monopole, was pioneered by Alvarez and co-workers (Alvarez et al. 1971) and later advanced by Cabrera (1982), whose first detector recorded a current jump consistent with a monopole on 14 February 1982, corresponding at that time to a flux of 6×10^{-10} cm^{-2} sr^{-1} sec^{-1}. Since then, he and other investigators have improved the sensitivity of their searches by several factors of 10 without making another detection, so the initial event is considered to have been spurious. The existence of the galactic magnetic field, which would be dissipated by the presence of monopoles, places an independent upper limit on the monopole flux of about 10^{-16} cm^{-2} sr^{-1} sec^{-1} (Parker 1979). In fact, there is no reason to think that the monopole flux is even as high as this—there may be none at all.

The range of potential applications of magnetic monopoles, should they exist, is quite extensive. A simple, though highly practical, example is the following. Imagine placing an electric charge on one end of a rigid nonconducting rod and a magnetic charge on the other end. Since the magnetic monopole's field is everywhere radial, any motion of the electron other than in a direction along the rod would subject it to $\mathbf{v} \times \mathbf{B}$ forces that are always perpendicular to its velocity and perpendicular to the rod (see Equation [1.3]). At best, the electron would execute circular motion in a plane perpendicular to the local $\mathbf{B}$ (and hence perpendicular to the rod). If kept in a low temperature region, where the equipartition energy of the electron is small compared to its electromagnetic potential energy so that its gyration radius is much smaller than the length of the rod, such a system would be prevented from rotating with an angular momentum vector perpendicular to its length, and would thus make a very effective gyroscope with no moving parts.

We will "grudgingly" assume that the experimental evidence does not yet support the existence of magnetic monopoles, and that a finite volume cannot therefore contain a source or sink of magnetic flux. In any case, even if there are a few monopoles, it is clear that they are so rare that they can be ignored for ordinary purposes. We adopt the view that every magnetic field line that enters a closed surface must exit somewhere else on that surface. Mathematically, this fact is expressed as

---

[3]For a review of experimental techniques and flux limits from induction and ionization monopole searches, see Groom (1985) and the resource letter of Goldhaber and Trower (1990). An earlier discussion of the relevance of monopoles to particle physics was produced by Schwinger (1969).

$$\oint_S \mathbf{B} \cdot d\mathbf{a} = 0 \, . \tag{1.13}$$

As was the case for the electric field, we would like to have a law that deals with the flux threading the surface enclosing an infinitesimal volume element, so that with our definition of divergence, we will modify Equation (1.13) to arrive at the second of Maxwell's equations:

$$\boxed{\operatorname{div} \mathbf{B} \equiv \vec{\nabla} \cdot \mathbf{B} = 0} \, . \tag{1.14}$$

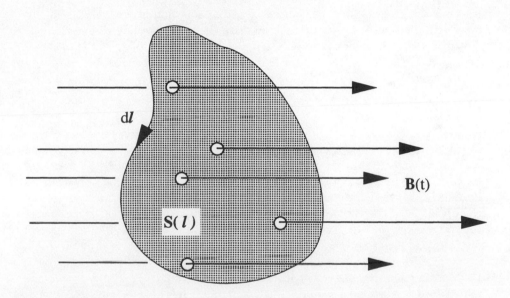

Figure 1.2 Contour integration of **E** around a closed path enclosing an area $S(l)$ and threaded by a changing magnetic flux.

Now a property of the electrostatic field is that it is conservative, meaning that the work performed by electrostatic forces is zero when a charge moves around a closed path, i.e., $\oint \mathbf{E} \cdot d\mathbf{l} = 0$ (see Figure 1.2). Therefore, we can say that the electric field is derivable from a *scalar* function Φ (the potential), which depends only on the spatial coordinates of the point where the field is to be calculated. The reason that **E** is a vector whereas Φ is a scalar is that the potential Φ can change in three independent ways corresponding to the three independent spatial directions. The field components are a measure of the rate at which Φ changes in each direction, so we need three different components to describe $\mathbf{E}$. Thus,

$$\mathbf{E} = -\operatorname{grad} \Phi \equiv -\vec{\nabla} \Phi \ . \tag{1.15}$$

Faraday's contribution was to show experimentally that $\oint \mathbf{E} \cdot d\mathbf{l} \neq 0$ when the closed path is linked by a changing magnetic flux Φ_B. In that case,

$$\oint \mathbf{E} \cdot d\mathbf{l} = -\frac{1}{c} \frac{d\Phi_B}{dt} \equiv -\frac{d}{dt} \int_{S(l)} \frac{1}{c} \mathbf{B} \cdot d\mathbf{a} \ . \tag{1.16}$$

As with the integral form of Gauss's law, this equation is not very useful for most situations. We are again motivated to define a new operation, viz., the *curl* of a vector $\mathbf{E}$, which is to be the limiting *circulation* of $\mathbf{E}$ as the enclosed area S vanishes. Thus, the curl of $\mathbf{E}$ in a direction $\hat{S}$ is given as

$$(\operatorname{curl} \mathbf{E})_{\hat{S}} = \lim_{S \to 0} \left\{ \frac{1}{S} \oint \mathbf{E} \cdot d\mathbf{l} \right\} \ , \tag{1.17}$$

where $\hat{S}$ is a unit vector oriented normal to the surface S. As was the case earlier in the definition of divergence, the limit in this expression must be taken with care. It is understood that S is shrunk to the point $\mathbf{x}_0$ at which the curl is to be calculated in such a way as to at each stage always contain $\mathbf{x}_0$ within its boundary, and that the contour of the integral always denotes this boundary of S, so that both the contour and S change in the limit process. Curl measures the circulation $\oint \mathbf{E} \cdot d\mathbf{l}$ per unit area so, for example, curl $\mathbf{E} = 0$ in the case of a static, and hence conservative, field.

In a Cartesian coordinate system, the calculation of curl $\mathbf{E}$ would proceed as follows. Let us set up the surface so that it lies in the $x - y$ plane, with $\hat{S}$ pointing in the $\hat{z}$ direction (see Figure 1.3). The circulation Γ around the surface element $dx\,dy$ is

$$\Gamma = \left(\frac{\partial E_y}{\partial x} - \frac{\partial E_x}{\partial y} \right) dx\,dy \ . \tag{1.18}$$

Thus, putting $S = dx\,dy$, we get

$$(\operatorname{curl} \mathbf{E})_{\hat{z}} = \frac{\partial E_y}{\partial x} - \frac{\partial E_x}{\partial y} \ . \tag{1.19}$$

The other components follow from $x - y - z$ permutations. Therefore,

$$\operatorname{curl} \mathbf{E} \equiv \vec{\nabla} \times \mathbf{E} \ . \tag{1.20}$$

Realizing that for a continuous field **B**

$$\lim_{S \to 0} \left\{ \frac{1}{S} \int_S \mathbf{B} \cdot d\mathbf{a} \right\} = \mathbf{B} \cdot \hat{S} = B_z \ , \tag{1.21}$$

we arrive at the third Maxwell equation,

$$\boxed{\operatorname{curl} \mathbf{E} \equiv \vec{\nabla} \times \mathbf{E} = -\partial \mathbf{B}/\partial \, ct} \ . \tag{1.22}$$

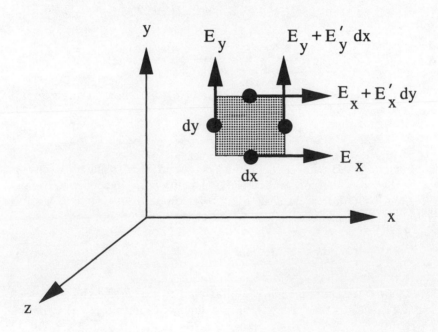

Figure 1.3 The curl of the electric field **E** in the xy-plane of a Cartesian coordinate system. In this figure, $E'_x \equiv \partial E_x/\partial y$ and $E'_y \equiv \partial E_y/\partial x$.

Note that the derivative appearing in Equation (1.16) is the *total* time derivative which accounts for the changes in flux through the loop due to both the local variation in **B** at a fixed point and the differences in **B** encountered by the loop as it moves through a region of nonuniform magnetic field. This derivative, often referred to as the *convective* derivative, is given by the chain rule of differentiation, $d/dt = \partial/\partial t + (dx/dt)(\partial/\partial x) + (dy/dt)(\partial/\partial y) + (dz/dt)(\partial/\partial z)$, or in more compact form, $d/dt = \partial/\partial t + \mathbf{v} \cdot \vec{\nabla}$. Since we are here considering a stationary loop in the frame where $\mathbf{B}$ is measured, we have $\mathbf{v} = 0$ and only the partial time derivative appears in Equation (1.22). The behavior of the fields in the context of moving loops (or frames) will be featured in Chapter 5.

Just as Gauss's law for time-independent conditions was deduced from Coulomb's law for the electric field due to a charge Q, the time-independent form of Ampère's law (the fourth Maxwell equation) is deduced from (and ultimately finds its justification in) the Biot-Savart rule for the magnetic field due to a *moving* charge Q_i located at $\mathbf{x}_i$, which gives rise to a current density $J_i(\mathbf{x}) \equiv Q_i v_i(\mathbf{x})\, \delta^3(\mathbf{x} - \mathbf{x}_i)$. From Equation (1.5), it is evident that for a time-independent current

$$\mathbf{B} = \frac{1}{c} \int \frac{I_i\, d\mathbf{l} \times \hat{r}}{r^2} \ , \tag{1.23}$$

where $I_i \equiv \int_S \mathbf{J}_i \cdot d\mathbf{a}$. For a constant electric field (see below), Ampère's law simply states that

$$\oint \mathbf{B} \cdot d\mathbf{l} = \frac{4\pi}{c} \sum_i I_i \ , \tag{1.24}$$

i.e., the *circulation* of $\mathbf{B}$ (the left-hand side) is proportional to the total current enclosed by the loop. The constant of proportionality is again chosen to comply with the empirically derived Equation (1.23) for the magnetic field. If we now let $\mathbf{J}$ represent the total current density, so that

$$\int_S \mathbf{J} \cdot d\mathbf{a} = \sum_i I_i \ , \tag{1.25}$$

and if we follow the same procedure as that used to derive the third Maxwell equation (1.22), we can take the infinitesimal limit of Equation (1.24) to obtain

$$\text{curl } \mathbf{B} \equiv \vec{\nabla} \times \mathbf{B} = \frac{4\pi}{c} \mathbf{J} \ . \tag{1.26}$$

However, this equation is not complete as it stands, since it does not appear to apply to time-dependent situations—it violates another very important experimental fact: charge appears to be absolutely conserved. To see how this comes about, let us digress for a moment to understand what in fact is required for the conservation of charge. Suppose a volume V encloses a total charge Q with charge density $\rho(\mathbf{x}, t)$:

$$Q = \int_V \rho(\mathbf{x}, t)\, d^3x \ . \tag{1.27}$$

Suppose now that the charge distribution has a velocity $\mathbf{v}(\mathbf{x}, t)$. Then, as the volume changes, or the surface area S moves, the density can change too and

$$\frac{dQ}{dt} = \lim_{\delta t \to 0} \frac{1}{\delta t} \left\{ \int_{V'} \rho(\mathbf{x}, t + \delta t) \, d^3x - \int_V \rho(\mathbf{x}, t) \, d^3x \right\}, \qquad (1.28)$$

where V' is the volume at time $t + \delta t$. Thus, Q may change not only as a result of a variation in ρ, but also in response to a change in V:

$$\frac{dQ}{dt} = \lim_{\delta t \to 0} \frac{1}{\delta t} \left\{ \int_V \rho(\mathbf{x}, t + \delta t) \, d^3x + \int_{\Delta V} \rho(\mathbf{x}, t + \delta t) \, d^3x - \int_V \rho(\mathbf{x}, t) \, d^3x \right\}$$

$$= \lim_{\delta t \to 0} \left\{ \frac{1}{\delta t} \int_V [\rho(\mathbf{x}, t + \delta t) - \rho(\mathbf{x}, t)] \, d^3x + \frac{1}{\delta t} \int_{\Delta V} \rho(\mathbf{x}, t + \delta t) \, d^3x \right\}, \quad (1.29)$$

where $\Delta V \equiv V' - V$. But for a small volume change ΔV, we can put $\int_{\Delta V} d^3x = \int_S \mathbf{v} \cdot \mathbf{da} \, \delta t$ (see Figure 1.4), and

$$\frac{dQ}{dt} = \int_V \frac{\partial \rho}{\partial t} \, d^3x + \lim_{\delta t \to 0} \frac{1}{\delta t} \int_S \rho(\mathbf{x}, t + \delta t) \, (\mathbf{v} \cdot \mathbf{da} \, \delta t)$$

$$= \int_V \frac{\partial \rho}{\partial t} \, d^3x + \int_S \rho(\mathbf{x}, t) \mathbf{v} \cdot \mathbf{da} . \qquad (1.30)$$

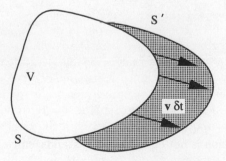

Figure 1.4 The shift in volume V containing the charge used to infer the continuity equation. During a time δt, the enclosing surface S moves a distance $\mathbf{v}\delta t$.

If charge is to be conserved, we must have

$$\lim_{V \to 0} \left\{ \frac{1}{V} \frac{dQ}{dt} \right\} = 0 \,, \tag{1.31}$$

from which we immediately get the so-called *conservation of charge equation*:

$$\boxed{\partial \rho / \partial t + \vec{\nabla} \cdot \mathbf{J} = 0} \,, \tag{1.32}$$

where $\mathbf{J}(\mathbf{x}) \equiv \rho(\mathbf{x}, t)\, \mathbf{v}(\mathbf{x}, t)$.

Returning now to Equation (1.26), we can see why this form of Ampère's law is incomplete, for taking the divergence of both sides, we get

$$\vec{\nabla} \cdot (\vec{\nabla} \times \mathbf{B}) = 0 = \frac{4\pi}{c} \vec{\nabla} \cdot \mathbf{J} \,. \tag{1.33}$$

This is clearly valid only when $\partial \rho / \partial t = 0$, otherwise $\vec{\nabla} \cdot \mathbf{J}$ cannot be zero. Maxwell's contribution was to add a term which correctly takes into account the missing $\partial \rho / \partial t$. Direct inspection of Equation (1.12) shows that we need a second term $\partial \mathbf{E} / \partial ct$ on the right-hand side of (1.26), and so the correct form of the fourth Maxwell equation is

$$\boxed{\mathrm{curl}\, \mathbf{B} \equiv \vec{\nabla} \times \mathbf{B} = 4\pi \mathbf{J}/c + \partial \mathbf{E} / \partial ct} \,. \tag{1.34}$$

It is useful to note that the presence of the second term on the right-hand side of this equation guarantees that in a time-dependent setting the current $\mathbf{J}$ is always brought into compliance with Ampère's law. Because the current depends on the mechanics of the particle motion, it might appear that $\mathbf{B}$ can change so fast that $\mathbf{J}$ may not always be adequate to account for the curl of $\mathbf{B}$. However, the ensuing rapid change in $\mathbf{E}$ in the direction of $\vec{\nabla} \times \mathbf{B}$ always forces $\mathbf{J}$ to attain the appropriate value as long as there is a significant number of movable charges present that can be accelerated by $\mathbf{E}$.

Maxwell's equations constitute eight scalar field equations in six unknowns, so they cannot all be independent. In a typical situation, ρ and $\mathbf{J}$ are specified in a well-defined region of spacetime and $\mathbf{E}$ and $\mathbf{B}$ are to be determined. In many cases, the equations may be simplified further by writing them in terms of potentials. We have already encountered the scalar potential Φ for the electric field. Since div $\mathbf{B} = 0$, $\mathbf{B}$ may itself be written as a potential by using the fact that the divergence of a curl is zero. (Although it's easy to show this mathematically, this point also makes intuitive sense because obviously a curl does not generate a net flux through a closed surface.) It is therefore expected that a vector field $\mathbf{A}$ exists such that

$$\mathbf{B} = \operatorname{curl} \mathbf{A} = \vec{\nabla} \times \mathbf{A} . \tag{1.35}$$

With these definitions, it is straightforward to see that the second Maxwell equation simply defines the vector potential according to Equation (1.35), and the third equation (also known as Faraday's law) can be converted into an expression for $\mathbf{E}$ in the presence of time-varying potentials:

$$\operatorname{curl} \mathbf{E} = -\frac{1}{c} \frac{\partial}{\partial t} \operatorname{curl} \mathbf{A}$$

$$\text{or} \quad \operatorname{curl} \left(\mathbf{E} + \frac{1}{c} \frac{\partial \mathbf{A}}{\partial t} \right) = 0 . \tag{1.36}$$

Since the quantity in parentheses must be the gradient of a scalar function, it is evident that

$$\mathbf{E} = -\frac{1}{c} \frac{\partial \mathbf{A}}{\partial t} - \vec{\nabla} \Phi . \tag{1.37}$$

Thus, the first Maxwell equation (1.12) is transformed into

$$\vec{\nabla}^2 \Phi + \frac{1}{c} \frac{\partial}{\partial t} (\vec{\nabla} \cdot \mathbf{A}) = -4\pi\rho , \tag{1.38}$$

which reduces to Poisson's equation for the scalar potential when the fields are time-independent. Finally, Ampère's law (Equation [1.34]) becomes

$$\vec{\nabla} (\vec{\nabla} \cdot \mathbf{A}) - \vec{\nabla}^2 \mathbf{A} + \frac{1}{c^2} \frac{\partial^2 \mathbf{A}}{\partial t^2} + \frac{1}{c} \frac{\partial}{\partial t} \vec{\nabla}\Phi = \frac{4\pi}{c} \mathbf{J} . \tag{1.39}$$

These reductions leave us with four equations in four unknowns: the scalar potential Φ and the three components of the vector potential $\mathbf{A}$. We must note, however, that although we have simplified the set of equations (from eight to four), we have not reduced the amount of information required to solve for $\mathbf{E}$ and $\mathbf{B}$. Each of the equations in (1.38) and (1.39) is second order and therefore the set as a whole is eighth order, requiring two integral conditions for a complete solution (one for each set of integrations). Once these potentials are determined, the fields $\mathbf{E}$ and $\mathbf{B}$ can be calculated from Equations (1.35) and (1.37).

1.2   MAXWELL'S EQUATIONS IN MATTER

The equations we have derived describe the electric, $\mathbf{E}$, and magnetic, $\mathbf{B}$, fields everywhere in spacetime, provided all the sources ρ and $\mathbf{J}$ are completely specified. However, for macroscopic aggregates of matter, solving the equations is almost impossible because of the *complexity* of the currents and charge densities. Instead, it is often convenient in such circumstances to employ an averaging process that allows us to identify the *macroscopic* fields. In matter, the sources may be decomposed according to

$$\rho = \rho_{\text{free}} + \rho_{\text{bound}} , \qquad (1.40)$$

and

$$\mathbf{J} = \mathbf{J}_{\text{free}} + \mathbf{J}_{\text{bound}} , \qquad (1.41)$$

where ρ_{free} and $\mathbf{J}_{\text{free}}$ are the sources that give rise to $\mathbf{E}$ and $\mathbf{B}$ in vacuum, and ρ_{bound} and $\mathbf{J}_{\text{bound}}$ represent the response of the medium to the presence of the fields. The *macroscopic* Maxwell equations are complicated by the fact that ρ_{bound} and $\mathbf{J}_{\text{bound}}$ are themselves sources of $\mathbf{E}$ and $\mathbf{B}$.

In matter, the four Maxwell equations can be written

$$\vec{\nabla} \cdot \mathbf{E} = 4\pi \rho_{\text{free}} + 4\pi \rho_{\text{bound}} , \qquad (1.42)$$

$$\vec{\nabla} \cdot \mathbf{B} = 0 , \qquad (1.43)$$

$$\vec{\nabla} \times \mathbf{E} + \frac{1}{c}\frac{\partial \mathbf{B}}{\partial t} = 0 , \qquad (1.44)$$

and

$$\vec{\nabla} \times \mathbf{B} - \frac{1}{c}\frac{\partial \mathbf{E}}{\partial t} = \frac{4\pi}{c}\mathbf{J}_{\text{free}} + \frac{4\pi}{c}\mathbf{J}_{\text{bound}} . \qquad (1.45)$$

The applied electric field $\mathbf{E}$ induces a separation of charges in matter so that the main effect of ρ_{bound} is to produce a "polarization" field (i.e., a field that locally points from the accumulation of positive charges to the segregated negative charges). To emphasize the fact that ρ_{bound} is giving rise to an electric field, we write an expression for it analogous to Equation (1.12), though replacing $\mathbf{E}$ with the *polarization* vector, $\mathbf{P}$, which represents the electric field produced by this charge separation:

$$\rho_{\text{bound}} \equiv -\vec{\nabla} \cdot \mathbf{P} . \qquad (1.46)$$

Thus, the first of Maxwell's laws can be reformulated as

$$\boxed{\vec{\nabla} \cdot \mathbf{D} = 4\pi \rho_{\text{free}}} \ , \tag{1.47}$$

where $\mathbf{D}$ is the dielectric displacement vector (i.e., the electric field measured inside the medium). In terms of the dipole moment of the material, $\mathbf{P}$, per unit volume, we have

$$\mathbf{D} = \mathbf{E} + 4\pi \mathbf{P} \ . \tag{1.48}$$

A similar procedure may be followed for the fourth Maxwell equation,[4] which results in the expression

$$\boxed{\vec{\nabla} \times \mathbf{H} - \partial \mathbf{D}/\partial ct = 4\pi \mathbf{J}_{\text{free}}/c} \ . \tag{1.49}$$

The magnetic field $\mathbf{H}$ is given explicitly in terms of the magnetic induction $\mathbf{B}$ and the macroscopically averaged magnetic dipole moment of the medium, $\mathbf{M}$, per unit volume:

$$\mathbf{H} \equiv \mathbf{B} - 4\pi \mathbf{M} \ . \tag{1.50}$$

Notice that because the homogeneous Equations (1.14) and (1.22) do not directly involve the sources ρ and $\mathbf{J}$, they do not change in the presence of matter. Henceforth, we shall drop the subscript "free," and ρ and $\mathbf{J}$ will always be understood to mean the free sources that give rise to the fields in vacuum. $\mathbf{E}$ and $\mathbf{B}$ are the fields everywhere, whereas $\mathbf{H}$ is the portion of the magnetic field produced by the external sources (i.e., $\mathbf{J}$ and $\vec{\nabla} \times \mathbf{E}$), and $\mathbf{D}$ is the electric field plus the polarization. The latter two are most usefully employed in the presence of matter.

Let us now try to understand the nature of the fields $\mathbf{D}$ and $\mathbf{H}$. Unlike their counterparts in vacuum, the Maxwell equations in matter represent eight relations in 12 unknowns. Thus, the equations cannot be solved exactly until we know the *constitutive* relations between $\mathbf{D}$, $\mathbf{H}$, and $\mathbf{E}$ and $\mathbf{B}$:

$$\mathbf{D} = \mathbf{D}(\mathbf{E}, \mathbf{B}) \ , \tag{1.51}$$

and

$$\mathbf{H} = \mathbf{H}(\mathbf{E}, \mathbf{B}) \ . \tag{1.52}$$

Under most conditions (the presence of a strong $\mathbf{E}$ or $\mathbf{B}$ represents an exception), the electric and magnetic polarization is proportional to the magnitude

---

[4]For example, see Chapter 6 in Jackson (2000).

of the applied field. When this is true, the medium is said to be linear, and in that case

$$D_i = \sum_j \varepsilon_{ij} E_j \,, \tag{1.53}$$

and

$$H_i = \sum_j \mu'_{ij} B_j \,, \tag{1.54}$$

where ε_{ij} and μ'_{ij} are the electric permittivity (or dielectric) tensor and the inverse magnetic permeability tensor, respectively. In the simplest cases, the linear response is isotropic and the ε_{ij} and μ'_{ij} tensors are diagonal, with all three elements equal. Thus, for linear, isotropic media,

$$\mathbf{D} = \varepsilon \, \mathbf{E} \,, \tag{1.55}$$

and

$$\mathbf{H} = \mu' \, \mathbf{B} \,. \tag{1.56}$$

However, ferromagnetic substances, e.g., iron, are only very poorly described by these linear relations because of hysteresis effects. Paramagnetic and diamagnetic substances are better represented in relatively weak fields.

We end this subsection by remarking once again that the description of classical electrodynamics is in terms of fields that are continuous functions of the coordinates of spacetime points. The formulation of this theory is therefore of necessity in terms of partial differential equations with many variables. This is in contrast with particle dynamics, for example, where the coordinates of the particles satisfy ordinary differential equations (ODEs). To solve an ODE for the trajectory of a particle, we simply need initial values of the position $\mathbf{x}(0)$ and momentum $\mathbf{p}(0)$. But for partial differential equations, the role of the boundary is much more fundamental, and we must therefore specify the *boundary conditions* as well as the initial state.

It is also appropriate to wonder how the quantum of charge enters into a classical description such as we have here. Our treatment involves the macroscopic fields, for which the equations are independent of the nature of ρ, except that ρ must satisfy the continuity (or charge conservation) Equation (1.32). On a microscopic level, we would write

$$\rho(\mathbf{x}) = \sum_i Q_i \, \delta^3(\mathbf{x} - \mathbf{x}_i) \,, \tag{1.57}$$

and

$$\mathbf{J}(\mathbf{x}) = \sum_i Q_i \, \mathbf{v}_i \, \delta^3(\mathbf{x} - \mathbf{x}_i) \,. \tag{1.58}$$

In principle, a microscopic formulation for the electromagnetic field could thus also be handled adequately, except that divergences resulting from self-energies require the special renormalization techniques of quantum electrodynamics.[5]

1.3 THE MATHEMATICAL STRUCTURE OF ELECTRODYNAMICS

The Maxwell equations, together with the necessary dynamical equations for the charges in the system (e.g., the Lorentz force equation), can in many instances be simplified considerably. Several classes of problems permit us to employ approximation schemes that often reduce the complete set of equations to as few as two, sometimes even just one (see § 1.3.1 below). It is useful, therefore, to discuss the mathematical structure of electrodynamics within the principal categories for which these approximations are valid. We will thus begin by laying out four of these subdivisions, and then discuss the mathematical apparatus needed to solve the four Maxwell equations within this framework.

1.3.1 Electrostatic Phenomena

When the magnetic field $\mathbf{H}$ and magnetic induction $\mathbf{B}$ are zero, Maxwell's equations reduce to

$$\operatorname{div} \mathbf{D} = \vec{\nabla} \cdot \mathbf{D} = 4\pi\rho\,, \tag{1.59}$$

and

$$\operatorname{curl} \mathbf{E} = \vec{\nabla} \times \mathbf{E} = 0\,. \tag{1.60}$$

The only constitutive relation we need is (1.51), which for linear, isotropic media is simply given by (1.55). Thus, Gauss's law (1.59) may be written as

$$\vec{\nabla} \cdot (\varepsilon\,\mathbf{E}) = 4\pi\rho\,, \tag{1.61}$$

or

$$\varepsilon\,\vec{\nabla} \cdot \mathbf{E} + (\vec{\nabla}\varepsilon) \cdot \mathbf{E} = 4\pi\rho\,. \tag{1.62}$$

As we have already seen (Equation [1.15]), Faraday's law is simply a statement of the conservative nature of an electrostatic field. When combined with Equation (1.62), it results in the expression

$$-\varepsilon\,\vec{\nabla}^2\Phi - (\vec{\nabla}\varepsilon) \cdot (\vec{\nabla}\Phi) = 4\pi\rho\,, \tag{1.63}$$

---

[5]A very readable account of the physics of quantization of a classical field may be found in Sakurai (1973).

which is an *elliptic* partial differential equation (see below). Thus, we are left with one equation in one unknown. If in addition $\varepsilon = $ constant (e.g., in vacuum, where $\varepsilon = 1$), then Equation (1.63) reduces to Poisson's equation for the scalar potential Φ:

$$\boxed{\vec{\nabla}^2 \Phi = -4\pi\rho/\varepsilon}\ . \tag{1.64}$$

In a sourceless region of spacetime, this reduces to the even simpler Laplace equation, viz.,

$$\vec{\nabla}^2 \Phi = 0\ . \tag{1.65}$$

1.3.2　Magnetostatic Phenomena

When the electric field **E** and electric displacement vector **D** are zero, Maxwell's equations reduce to the set

$$\mathrm{div}\ \mathbf{B} = \vec{\nabla} \cdot \mathbf{B} = 0\ , \tag{1.66}$$

and

$$\mathrm{curl}\ \mathbf{H} = \vec{\nabla} \times \mathbf{H} = \frac{4\pi}{c}\ \mathbf{J}\ . \tag{1.67}$$

Here, the constitutive relation is just (1.52), which for a linear, isotropic medium may be written as

$$\mathbf{H} = \frac{1}{\mu}\ \mathbf{B}\ , \tag{1.68}$$

with $\mu \equiv 1/\mu'$ defined to be the magnetic permeability. Thus,

$$\vec{\nabla} \times \left(\frac{\mathbf{B}}{\mu}\right) = \left(\vec{\nabla}\frac{1}{\mu}\right) \times \mathbf{B} + \frac{1}{\mu}\left(\vec{\nabla} \times \mathbf{B}\right) = \frac{4\pi}{c}\ \mathbf{J}\ . \tag{1.69}$$

If in addition $\mu = $ constant, then

$$\vec{\nabla} \times \mathbf{B} = \frac{4\pi\mu}{c}\ \mathbf{J}\ . \tag{1.70}$$

As we have already seen, the fact that div **B** = 0 implies that **B** is the curl of another vector field $\mathbf{A}(\mathbf{x})$ (see Equation [1.35]). This brings us to what at first appears to be an innocuous property of the vector potential, but which was in fact the origin of subsequent Gauge theories (e.g., the Yang-Mills theory)[6]

---

[6]See the seminal paper by Yang and Mills (1954).

and all theories of supersymmetry built upon them. Since $\vec{\nabla} \times (\vec{\nabla}\psi) = 0$ for any scalar field ψ, the vector potential can be freely transformed by adding the gradient of ψ without affecting the magnetic induction:

$$\mathbf{A} \to \mathbf{A} + \vec{\nabla}\psi. \tag{1.71}$$

Such a *gauge transformation* permits $\vec{\nabla} \cdot \mathbf{A}$ to have any convenient functional form we wish. For example, the expression that results from a direct substitution of (1.35) into (1.70), i.e.,

$$\vec{\nabla} \times (\vec{\nabla} \times \mathbf{A}) = \vec{\nabla}(\vec{\nabla} \cdot \mathbf{A}) - \vec{\nabla}^2 \mathbf{A} = \frac{4\pi\mu}{c} \mathbf{J}, \tag{1.72}$$

can be greatly simplified with the choice of gauge

$$\vec{\nabla} \cdot \mathbf{A} = 0 \tag{1.73}$$

(the so-called *Coulomb gauge*—see § 6.4), for which the components of $\mathbf{A}$ must then satisfy the Poisson equation:

$$\boxed{\vec{\nabla}^2 A_i = -4\pi\mu J_i/c \qquad (i = 1, 2, 3)}. \tag{1.74}$$

It is interesting to note that the condition $\vec{\nabla} \cdot \mathbf{A} = 0$ reduces to the equation $\nabla^2\psi = 0$ (as we shall see later), so that if there are no sources at infinity, ψ must be a constant.

1.3.3   Wave Phenomena

In a region where there are no free sources ($\rho = 0$ and $\mathbf{J} = 0$), Maxwell's equations reduce to

$$\vec{\nabla} \cdot \mathbf{D} = 0, \tag{1.75}$$

$$\vec{\nabla} \cdot \mathbf{B} = 0, \tag{1.76}$$

$$\vec{\nabla} \times \mathbf{E} = -\frac{1}{c}\frac{\partial \mathbf{B}}{\partial t}, \tag{1.77}$$

$$\vec{\nabla} \times \mathbf{H} = \frac{1}{c}\frac{\partial \mathbf{D}}{\partial t}. \tag{1.78}$$

In order to proceed from here, we must also know the constitutive relations (1.51) and (1.52). Let us assume for the sake of simplicity that $\mathbf{D}$ and $\mathbf{H}$ are given by Equations (1.55) and (1.56), respectively, and that ε and $\mu' = 1/\mu$ are constants, so that all derivatives of ε and μ are zero. Then,

$$\vec{\nabla} \cdot \mathbf{E} = 0 , \tag{1.79}$$

$$\vec{\nabla} \cdot \mathbf{B} = 0 , \tag{1.80}$$

$$\vec{\nabla} \times \mathbf{E} = -\frac{1}{c} \frac{\partial \mathbf{B}}{\partial t} , \tag{1.81}$$

and

$$\vec{\nabla} \times \mathbf{B} = \frac{\mu \varepsilon}{c} \frac{\partial \mathbf{E}}{\partial t} . \tag{1.82}$$

These are simultaneous, first-order, differential equations. The idea is to eliminate as much as possible and to simplify. To this end, let us take the curl of (1.82) and add it to the time derivative of (1.81), which gives

$$\vec{\nabla} \times (\vec{\nabla} \times \mathbf{B}) = -\frac{\mu \varepsilon}{c^2} \frac{\partial^2 \mathbf{B}}{\partial t^2} . \tag{1.83}$$

But since $\vec{\nabla} \times (\vec{\nabla} \times \mathbf{B}) = \vec{\nabla}(\vec{\nabla} \cdot \mathbf{B}) - \vec{\nabla}^2 \mathbf{B}$ and $\vec{\nabla} \cdot \mathbf{B} = 0$, we get

$$\boxed{\vec{\nabla}^2 \mathbf{B} - \mu \varepsilon \, \partial^2 \mathbf{B}/\partial (ct)^2 = 0} . \tag{1.84}$$

This is a *hyperbolic* partial differential equation (see below), commonly known as the *wave equation*. One of its solutions may be written

$$\mathbf{B} = \mathbf{B}_0 \exp\{i (\mathbf{k} \cdot \mathbf{x} - \omega t)\} , \tag{1.85}$$

where $\mathbf{B}_0$ is a constant vector, and the angular frequency ω is related to the magnitude of the wave vector $\mathbf{k}$ by the relation

$$|\mathbf{k}| = \frac{\omega}{|\mathbf{v}|} . \tag{1.86}$$

Here, $|\mathbf{v}| \equiv c/\sqrt{\mu \varepsilon}$ is a constant of the medium, and it has the dimensions of a velocity. In vacuum, $\mu = \varepsilon = 1$ and $|\mathbf{v}| = c$. In matter, the solutions are still waves, but their velocity is smaller than c.

A similar reduction may be made for the electric field $\mathbf{E}$, which produces a second wave equation,

$$\boxed{\vec{\nabla}^2 \mathbf{E} - \mu \varepsilon \, \partial^2 \mathbf{E}/\partial (ct)^2 = 0} , \tag{1.87}$$

whose solution is

$$\mathbf{E} = \mathbf{E}_0 \exp\{i\,(\mathbf{k}\cdot\mathbf{x} - \omega\,t)\}\,. \tag{1.88}$$

The case of radial waves will be considered in § 8.2.

The presence of sources in the system (i.e., $\rho \neq 0$ and $\mathbf{J} \neq 0$) modifies the character of these waves considerably. In that case, a complete solution to the corresponding wave equations with sources on the right-hand side will include waves of the fields associated with the charges themselves, and these are not planar. But we are not fully equipped yet to discuss this situation. We will begin to explore the fields produced by distributions of charge in Chapter 2, and then we will derive the full framework for the solutions in Chapter 7, completing the process in § 8.2.

1.3.4   The General Case

In general, $\rho \neq 0$, $\mathbf{J} \neq 0$, and the fields are time dependent, as might occur, for example, when accelerated charges are radiating. For this situation, Maxwell's equations must be supplemented by a description of the particle dynamics, i.e.,

$$\mathbf{x}_i = \mathbf{x}_i(t)\,, \tag{1.89}$$

$$\rho(\mathbf{x}, t) = \sum_i Q_i\,\delta^3(\mathbf{x} - \mathbf{x}_i)\,, \tag{1.90}$$

$$\mathbf{J}(\mathbf{x}, t) = \sum_i Q_i\,\mathbf{v}_i\,\delta^3(\mathbf{x} - \mathbf{x}_i)\,, \tag{1.91}$$

and

$$m_i\,\frac{d^2\,\mathbf{x}_i}{d\,t^2} = \mathbf{F}_i = Q_i\left\{\mathbf{E}\,[\mathbf{x}_i(t)] + \frac{\mathbf{v}_i \times \mathbf{B}\,[\mathbf{x}_i(t)]}{c}\right\}\,, \tag{1.92}$$

where it is understood here that $\mathbf{E}\,[\mathbf{x}_i(t)]$ does not include the self-field due to the charge Q_i at $\mathbf{x}_i(t)$. We thus end up with a coupled electromagnetic-mechanical system.

1.3.5   The Mathematical Apparatus

It is evident from the above discussion of the four subdivisions of electrodynamics that we must deal with second-order, coupled, differential equations, whose coefficients are not always constant (e.g., when ε and μ vary within the medium).[7] The fields are functions of the generalized coordinates:

$$U = U(x_1, x_2, \ldots, x_n) \,, \tag{1.93}$$

where U represents either a scalar variable (e.g., Φ), or one of the components of a vector variable (e.g., A_i with $i = 1, 2, 3$). In many of the problems we will examine here, the differential equation for U looks like

$$LU = f \,, \tag{1.94}$$

where the "source" f is known (e.g., an electron whose charge density is given by Equation [1.90]), and the operator L has the form

$$L \equiv \sum_{i,j} a_{ij}(\mathbf{x}) \, \frac{\partial^2}{\partial x_i \, \partial x_j} + \sum_i b_i(\mathbf{x}) \, \frac{\partial}{\partial x_i} + c(\mathbf{x}) \,. \tag{1.95}$$

We shall encounter three possible types of partial differential equations: elliptic, hyperbolic, and parabolic.

Elliptic

Here, the symmetric matrix of coefficients $[a_{ij}]$ is positive definite for each $\mathbf{x}$ in the given domain. By definition, this means that all the eigenvalues of $[a_{ij}]$ are nonzero and have the same sign. The terminology arises by analogy with conic sections. For example, for an equation involving two independent variables x_1 and x_2, a very important role is played by the sign of the discriminant $a_{12}^2 - a_{11}a_{22}$. The operator L corresponding to this case is said to be of elliptic type when $a_{12}^2 < a_{11}a_{22}$. As an illustration, we might have the following equation written in Cartesian coordinates:

$$\vec{\nabla}^2 U = \frac{\partial^2 U}{\partial x^2} + \frac{\partial^2 U}{\partial y^2} + \frac{\partial^2 U}{\partial z^2} = f \,. \tag{1.96}$$

We shall see that for $f = 0$, the solution is unique when either U or $\vec{\nabla}U$ is specified on a closed bounding surface, but not both. Using U and $\vec{\nabla}U$ as boundary conditions overconstrains the system.

---

[7]There are several excellent descriptions of the types of differential equations and their associated boundary conditions. Among these are Sommerfeld (1949) and Courant and Hilbert (1962).

Hyperbolic

In this case, all the eigenvalues of the matrix $[a_{ij}]$ are nonzero and one of them has a different sign from the rest. For the general equation with two independent variables introduced above, the discriminant $a_{12}^2 - a_{11}a_{22}$ is here positive. An example is

$$\frac{\partial^2 U}{\partial x^2} - \frac{\partial^2 U}{\partial y^2} = f \ . \tag{1.97}$$

A second example is the wave equation we encountered in § 1.3.3 above. The latter has a *propagation* character in the sense that a profile in U is maintained at a spatial location that varies with time. The rate of change of this location is, of course, the wave velocity. In order to specify a unique solution here, we need to know not only $U(\mathbf{x}, 0)$ but $\partial U(\mathbf{x}, 0)/\partial t$ as well. In essence, this form of the equation adds the extra dimension of time into the problem. More generally, a hyperbolic equation can be solved uniquely only for a Cauchy problem, in which the field and its first derivative in one of the variables (e.g., time) are specified at some initial value of that variable everywhere on an open surface within the subspace defined by the remaining variables. The equation then describes how the field evolves forward in that chosen variable.

Parabolic

Finally, there is the situation where one of the eigenvalues of the matrix $[a_{ij}]$ is zero, and the others all have the same sign, i.e., where the matrix $[a_{ij}]$ is singular. For the two-dimensional equation described above, the discriminant $a_{12}^2 - a_{11}a_{22}$ is here exactly zero. An example is the equation

$$\frac{\partial U}{\partial x} + \frac{\partial^2 U}{\partial y^2} = f \ . \tag{1.98}$$

As we shall see in § 8.4, the magnetic field inside a weakly conducting medium "diffuses" outward according to the equation

$$\frac{\partial \mathbf{B}}{\partial t} = \frac{c^2}{4\pi\sigma} \vec{\nabla}^2 \mathbf{B} \ , \tag{1.99}$$

which is also of a parabolic form. Physical processes that can be described as such still have the "attribute" of propagation, but the transport now takes place at infinite speed. Diffusion is not a wave phenomenon, but rather, the rate of change of a field quantity at a given spacetime point is limited by a diffusion constant that depends on properties of the medium. Thus, for example, the diffusion of the magnetic field is controlled by the constant $D = c^2/4\pi\sigma$, which depends on the conductivity σ. The boundary conditions for parabolic equations are intermediate in nature between those of the elliptic and hyperbolic equations. For these equations, a unique solution requires specification of the field U on a Cauchy surface, i.e., the field must be known at the space boundaries of the given region. However, since the rate of diffusion is governed by the medium, no other conditions are necessary.

It is useful to classify the equations in this manner because the physical interpretation and boundary conditions are very different for the elliptic, hyperbolic, and parabolic forms. Remember that in electrodynamics the fields are defined throughout a designated region of spacetime, and therefore the boundary conditions affect the solutions everywhere within that region, unlike point particles for whom the trajectory in phase space depends only on the initial conditions.

2

TIME-INDEPENDENT FIELDS

2.1 ELECTROSTATICS

Henceforth, we shall consider the Maxwell equations as they pertain to electrodynamic fields associated with free charges and currents only, without the effects incurred by the presence of macroscopic aggregates of matter (see § 1.2 above). Thus, the electric permittivity tensor is the identity matrix (i.e., $\varepsilon = 1$). The benefit of this is that the essential physics is not lost in the reduced equations, but the manipulations are simpler. As we saw in § 1.3 above, all of the physics of electrostatics is contained within the deceptively simple equation

$$\vec{\nabla}^2 \, \Phi = -4\pi \, \rho \,, \tag{2.1}$$

or its equivalent form

$$\vec{\nabla} \cdot \mathbf{E} = 4\pi \, \rho \,. \tag{2.2}$$

This equation simply states that the divergence of the electric field, i.e., the outflux of the electric field per unit volume, is due to the presence of the source ρ. If we know ρ throughout a given region of space, these equations tell us exactly where the field originates and where it ends. Our task in this section is to learn the mathematical techniques for solving Equation (2.1), known as Poisson's equation. These techniques may be grouped under several headings, depending on which particular facet of the electric field they address first. Here, we will discuss three of the most commonly used categories, chosen for breadth (compare Methods 1 and 3) and connectivity (compare Methods 1 and 2).

2.1.1   Method 1: Guesses and Symmetries

The so-called *method of images* is probably the most commonly used technique by experienced practitioners. This method is particularly useful when there are one or more point charges in the presence of boundary surfaces. Under some circumstances, a small number of suitably placed *imaginary* charges outside the region of interest can simulate the boundary condition.

For example, suppose we need to calculate the potential Φ in a region where a charge q is near an infinite plane conductor maintained at zero potential. By symmetry, we expect that the boundary condition can be simulated by placing another charge $-q$ on the other side of the plane, so that the situation now looks like Figure 2.1.

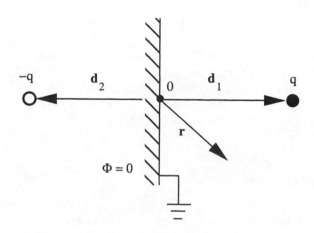

Figure 2.1   A charge q near a grounded, infinite conducting plane. An imaginary charge $-q$ positioned opposite q can mimic the effects of the boundary. The valid domain of solution is to the right of the boundary, where only "real" charges are permitted to exist.

The solution to Poisson's equation is then provided by the sum of the potentials of all the charges, but it is to be valid only in the domain of solution containing the real ones. In this example,

$$\Phi(\mathbf{r}) = \frac{q}{|\mathbf{r} - \mathbf{d}_1|} - \frac{q}{|\mathbf{r} - \mathbf{d}_2|} \cdot \tag{2.3}$$

It is not difficult to use this expression for Φ (a highly recommended exercise) to determine the surface charge distribution σ on the conductor. Remembering that at the interface Gauss's law gives $\mathbf{E} \cdot \hat{n} = 4\pi\,\sigma$, where $\hat{n}$ is the unit normal to the surface and $\mathbf{E}$ is taken to be zero inside the conductor, it is a trivial step to write $\vec{\nabla}\Phi \cdot \hat{n} = -4\pi\,\sigma$ or $\partial\Phi/\partial n = -4\pi\,\sigma$, where $\partial\Phi/\partial n$ is the normal derivative of Φ at the surface.

As a second application of this method using guesses and symmetries, let us consider what happens when we insert a grounded, spherical conductor of radius a in a given homogeneous electrostatic field $\mathbf{E}_0$ (Figure 2.2).

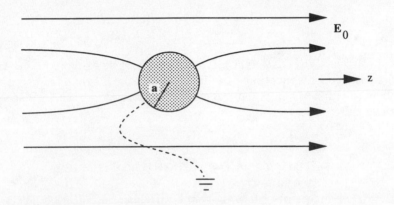

Figure 2.2 A grounded, conducting sphere embedded within a homogeneous electrostatic field $\mathbf{E}_0$.

What do we know? At $z \to \infty$, $\mathbf{E} \to \mathbf{E}_0$, the undisturbed electric field. There are no charges in this problem, so $\vec{\nabla}^2\Phi = 0$. The sphere is grounded, so $\Phi = 0$ at $r = a$. Also, since $\vec{\nabla}\Phi = -\mathbf{E}$ and $\mathbf{E} \to E_0\,\hat{k}$ at large z, we must have $\Phi = -E_0\,z$ as $z \to \infty$. We are led to the hypothesis that Φ should have the form $\Phi = \Phi_1 - E_0\,z$, with $\Phi_1 \to 0$ at large z. Since we have azimuthal symmetry, Φ should be independent of the azimuthal angle ϕ. Let us guess that $\Phi_1 = A\cos\theta\,r^\alpha$, the $\cos\theta$ coming from our suspicion that Φ_1 should depend on z $(= r\cos\theta)$. Then on the sphere,

$$A\cos\theta\,a^\alpha - E_0\cos\theta\,a = 0 , \tag{2.4}$$

which allows us to solve for the constant A:

$$A = E_0\,a^{1-\alpha} . \tag{2.5}$$

The other condition is $\vec{\nabla}^2 \Phi = 0$, which in spherical coordinates reduces to

$$\frac{1}{r^2} \frac{\partial}{\partial r} \left(r^2 \left[\alpha A \cos\theta \, r^{\alpha-1} - E_0 \cos\theta \right] \right) +$$

$$\frac{1}{r^2 \sin\theta} \frac{\partial}{\partial\theta} \left(\sin\theta \left[-A \sin\theta \, r^\alpha + E_0 \sin\theta \, r \right] \right) = 0 \,. \qquad (2.6)$$

There are two solutions to this equation, though only one matches the boundary condition. These correspond to the condition $\alpha(\alpha + 1) = 2$, for which $\alpha = -2$ and $\alpha = 1$. However, the latter does not give $\Phi_1 \to 0$ as $r \to \infty$, so we retain the former. The complete solution is evidently

$$\Phi = -E_0 \, r \, \cos\theta \left(1 - \frac{a^3}{r^3} \right) \,. \qquad (2.7)$$

The appearance of the $\cos\theta$ term will become clearer when we examine Method 3 for solving Poisson's equation, in which the potential is expanded using a series of orthonormal functions. A natural set of functions to use with this geometry is the Legendre polynomials, where the first-order term is indeed $\cos\theta$.

2.1.2   Method 2: The Green Function

The method of images described above is clearly most useful when the problem has a high degree of symmetry and a small number of charges, but it becomes intractable when the charge distribution and/or the boundary conditions are fairly complicated. The Green function technique that we consider here is a mathematical development of this basic method that in many ways "automatically" takes into account the effects of the boundary on the internal solution. But before we can develop the equations we need, we must first digress to derive the necessary tools.

Let us for the moment consider the field due to a point charge q. We know from Coulomb's law that

$$\mathbf{E} = \frac{q}{r^2} \, \hat{r} \,, \qquad (2.8)$$

and that therefore

$$\Phi = -\frac{q}{r} \,. \qquad (2.9)$$

It will prove useful to generalize this somewhat, by positioning the charge at $\mathbf{x}'$ instead of at the origin. (Our convention will be such that prime always denotes the source coordinates, whereas the unprimed variables pertain to the observation point and time.) Then,

$$\Phi(\mathbf{x}) = -\frac{q}{|\mathbf{x} - \mathbf{x}'|} \tag{2.10}$$

and

$$\mathbf{E}(\mathbf{x}) = -\vec{\nabla}\Phi = \frac{q(\mathbf{x} - \mathbf{x}')}{|\mathbf{x} - \mathbf{x}'|^3} \ . \tag{2.11}$$

For a general source distribution $\rho(\mathbf{x}')$, the potential is expected to be the sum over all increments of charge $d^3x'\,\rho(\mathbf{x}')$, i.e., it ought to look like the following:

$$\Phi(\mathbf{x}) = \int \frac{\rho(\mathbf{x}')\,d^3x'}{|\mathbf{x} - \mathbf{x}'|} \ . \tag{2.12}$$

We base this on our discussion in Chapter 1 regarding the fact that electric and magnetic fields do not themselves carry charge, so that the fields from different sources are not expected to interfere with each other. It is intuitively obvious, therefore, that this potential should satisfy Poisson's equation, right? But does it? Let's check:

$$\vec{\nabla}^2\Phi(\mathbf{x}) = \vec{\nabla}^2 \int \frac{\rho(\mathbf{x}')\,d^3x'}{|\mathbf{x} - \mathbf{x}'|} \ . \tag{2.13}$$

We must be careful here, because we have two variable position parameters: $\mathbf{x}$ and $\mathbf{x}'$. Remember that Poisson's equation tells us something about the field divergence at the point of observation, $\mathbf{x}$, so $\vec{\nabla}^2$ must operate on $\mathbf{x}$, not $\mathbf{x}'$. In that case, we can take $\vec{\nabla}^2$ inside the integral and get

$$\vec{\nabla}^2\Phi(\mathbf{x}) = \int \rho(\mathbf{x}')\,d^3x'\,\vec{\nabla}^2\left(\frac{1}{|\mathbf{x} - \mathbf{x}'|}\right) \ . \tag{2.14}$$

Now,

$$\vec{\nabla}^2\left(\frac{1}{|\mathbf{x} - \mathbf{x}'|}\right) = -\sum_{i=1}^{3} \partial_i \left[\frac{(x_i - x_i')}{|\mathbf{x} - \mathbf{x}'|^3}\right] = -\frac{3}{|\mathbf{x} - \mathbf{x}'|^3} + \frac{3}{|\mathbf{x} - \mathbf{x}'|^3} \ , \tag{2.15}$$

where following the usual convention, $\partial_i \equiv \partial/\partial x_i$. Of course, this makes sense as long as $\mathbf{x} \neq \mathbf{x}'$. In that case, $\vec{\nabla}^2(|\mathbf{x} - \mathbf{x}'|^{-1}) = 0$, as expected, since the divergence of electric flux should be zero from a region of space where there are no charges. However, at $\mathbf{x} = \mathbf{x}'$ the expression is not yet defined. This suggests that Equation (2.14) should be written

$$\vec{\nabla}^2\Phi(\mathbf{x}) = \rho(\mathbf{x}')|_{\mathbf{x}'=\mathbf{x}} \lim_{a\to 0} \int \vec{\nabla}^2\left(\frac{1}{\sqrt{r^2 + a^2}}\right)\,d^3x' \ , \tag{2.16}$$

where $r \equiv |\mathbf{x} - \mathbf{x}'|$ and we have taken ρ outside of the integral because the integrand is zero everywhere except possibly at $\mathbf{x} = \mathbf{x}'$. But $\vec{\nabla}^2([r^2 + a^2]^{-1/2}) = -3a^2(r^2 + a^2)^{-5/2}$, and so

$$\lim_{a \to 0} \int \vec{\nabla}^2 \left(\frac{1}{\sqrt{r^2 + a^2}} \right) d^3x' = \lim_{a \to 0} \left\{ -3a^2 \int_0^\infty 4\pi \, \frac{r^2 \, dr}{(r^2 + a^2)^{5/2}} \right\} = -4\pi \, . \tag{2.17}$$

We have thus recovered Poisson's equation, since we see that

$$\vec{\nabla}^2 \Phi(\mathbf{x}) = -4\pi \, \rho(\mathbf{x}')|_{\mathbf{x}'=\mathbf{x}} \, . \tag{2.18}$$

The singular nature of $\vec{\nabla}^2(|\mathbf{x} - \mathbf{x}'|^{-1})$ can best be expressed in terms of the Dirac delta function $\delta^3(\mathbf{x} - \mathbf{x}')$:[8]

$$\boxed{\vec{\nabla}^2(|\mathbf{x} - \mathbf{x}'|^{-1}) = -4\pi \, \delta^3(\mathbf{x} - \mathbf{x}')} \, . \tag{2.19}$$

By definition, $\int \delta^3(\mathbf{x} - \mathbf{x}') \, d^3x = 1$ if the integration volume contains the point $\mathbf{x} = \mathbf{x}'$ and is zero otherwise.

Thus, we have not only shown that the potential from Coulomb's law satisfies Poisson's equation, but we have established (through the solution to Poisson's equation) the important result that the potential from a distributed source is the superposition of the individual potentials from the constituent parcels of charge. In the infinitesimal limit, this is manifested as the integral appearing in Equation (2.12). In our discussion of the Green function technique, it will often be useful to visualize this property of the potential in terms of individual charges, not unlike the situation with the method of images we described above. More specifically, we will consider situations in which ρ is comprised of N discrete charges q_i positioned at $\mathbf{x}'_i$ for $i = 1, ..., N$, so that

$$\rho(\mathbf{x}') = \sum_{i=1}^N q_i \, \delta^3(\mathbf{x}' - \mathbf{x}'_i) \, . \tag{2.20}$$

In this case, the solution for the potential is simply a combination of terms proportional to $|\mathbf{x} - \mathbf{x}'_i|^{-1}$. The use of Green functions (constituting our Method 2) is merely an extension of this to problems that involve boundary conditions. We shall see that the potential from a given incremental source has the general

---

[8] A very useful summary of some of the properties of the Dirac delta function can be found in Appendix A of Davydov (1973).

form $\Phi \propto |\mathbf{x} - \mathbf{x}'_i|^{-1} + F(\mathbf{x}, \mathbf{x}'_i)$, where the function $F(\mathbf{x}, \mathbf{x}'_i)$ is included to satisfy the boundary constraint on the prespecified surface.

The way this works conceptually is that we mimic the effects of the boundary by pairing each parcel of charge (or each individual charge if we're dealing with a discrete system) inside the domain of solution with imaginary external charges. The function $F(\mathbf{x}, \mathbf{x}'_i)$ represents the contribution to the potential from these fictitious sources, but in a way that correctly accounts for the modification to Φ due to the presence of a boundary surface. But how do we do this? Let us take inventory and see what we have at our disposal. We know that no matter what the boundary condition is, the potential at $\mathbf{x}$ due to a charge δq at $\mathbf{x}'$ is $\delta q / |\mathbf{x} - \mathbf{x}'|$. If we knew what imaginary charges were needed to represent the boundary, we could simply solve the problem by using Equation (2.12) with an integration over all space, not just over the domain of solution, with ρ now representing the sum of the real and fictitious charge distributions. Unfortunately, we don't always know what the external charges are, but we do know what effect they must have on Φ or its derivative at the surface. What we need, therefore, is an expression that directly links these sources to the general form of the potential (or its derivative) evaluated at the boundary.

The need to link the volume integral in Equation (2.12) with a surface condition immediately suggests that we use the divergence theorem, which for a vector field $\mathbf{A}$ says that

$$\int_V \vec{\nabla}' \cdot \mathbf{A}(\mathbf{x}') \, d^3 x' = \oint_S \mathbf{A}(\mathbf{x}') \cdot \hat{n} \, da' , \qquad (2.21)$$

where V is the volume of integration, S is its enclosing surface, and $\hat{n}(\mathbf{x}')$ is a unit vector (externally) normal to the surface element da'. At this point, we can try several things, but our main goal is somehow to use this equation to relate the actual potential Φ on S to the incremental potential $\rho(\mathbf{x}') \, d^3 x' \, \psi(\mathbf{x}, \mathbf{x}')$. Here, we have defined the unit charge (template) potential $\psi \equiv |\mathbf{x} - \mathbf{x}'|^{-1}$, which will be useful in streamlining our derivations. It is to be thought of as a template in the sense that it represents the geometrical dependence of Φ, so that the total potential is a convolution of ψ with the given physical charge distribution ρ. Let's try putting $\mathbf{A}(\mathbf{x}') = \Phi(\mathbf{x}') \vec{\nabla}' \psi(\mathbf{x}, \mathbf{x}')$. We could also have tried $\mathbf{A}(\mathbf{x}') = \psi(\mathbf{x}, \mathbf{x}') \vec{\nabla}' \Phi(\mathbf{x}')$, which we will in fact need to do below. For the first definition of $\mathbf{A}$, we have

$$\vec{\nabla}' \cdot (\Phi \vec{\nabla}' \psi) = \Phi \vec{\nabla}'^2 \psi + \vec{\nabla}' \Phi \cdot \vec{\nabla}' \psi , \qquad (2.22)$$

and

$$\Phi \vec{\nabla}' \psi \cdot \hat{n} = \Phi \frac{\partial \psi}{\partial n} . \qquad (2.23)$$

Substituting these into Equation (2.21) produces the so-called Green first identity:

$$\int_V \left(\Phi \vec{\nabla}'^2 \psi + \vec{\nabla}' \Phi \cdot \vec{\nabla}' \psi \right) d^3 x' = \oint_S \Phi \frac{\partial \psi}{\partial n} \, da' \, . \tag{2.24}$$

We're almost there, but we still can't quite use this since we don't know how to evaluate $\vec{\nabla}' \psi$ and $\vec{\nabla}' \Phi$. We do, however, know what $\vec{\nabla}'^2 \psi$ is (see Equation [2.19]). The second term in the integrand on the left-hand side can conveniently be removed by subtracting Equation (2.24) from an analogous equation derived with Φ and ψ interchanged in the definition of $\mathbf{A}$. The result is *Green's theorem*:

$$\boxed{\int_V [\Phi \vec{\nabla}'^2 \psi - \psi \vec{\nabla}'^2 \Phi] \, d^3 x' = \oint_S [\Phi \partial \psi / \partial n - \psi \partial \Phi / \partial n] \, da'} \, . \tag{2.25}$$

(An alternative, though perhaps less physically motivated, derivation of this equation follows from integration of the left-hand side by parts.)

Now we can fully appreciate the relevance of Poisson's equation for a discrete charge (Equation [2.19]), for it allows us to extract the necessary relationship between Φ and the conditions on the boundary surface. Substituting for $\vec{\nabla}'^2 \psi$ in Equation (2.25), we get

$$\int_V \left[-4\pi \, \delta^3(\mathbf{x} - \mathbf{x}') \, \Phi(\mathbf{x}') + \frac{4\pi \, \rho(\mathbf{x}')}{|\mathbf{x} - \mathbf{x}'|} \right] d^3 x' =$$

$$\oint_S \left[\Phi \frac{\partial}{\partial n} \left(\frac{1}{|\mathbf{x} - \mathbf{x}'|} \right) - \frac{1}{|\mathbf{x} - \mathbf{x}'|} \frac{\partial \Phi}{\partial n} \right] da' \, . \tag{2.26}$$

The condition we've imposed all along is that the observation point $\mathbf{x}$ should lie within the volume V. Thus, integrating the Dirac delta function over all values of $\mathbf{x}'$ within V yields a nonzero result, and we get

$$\Phi(\mathbf{x}) \quad = \quad \int_V \frac{\rho(\mathbf{x}')}{|\mathbf{x} - \mathbf{x}'|} \, d^3 x'$$

$$+ \frac{1}{4\pi} \oint_S \left[\frac{1}{|\mathbf{x} - \mathbf{x}'|} \frac{\partial \Phi}{\partial n} - \Phi \frac{\partial}{\partial n} \left(\frac{1}{|\mathbf{x} - \mathbf{x}'|} \right) \right] da' \, . \tag{2.27}$$

Notice that this is just what we had before in Equation (2.12), but now with a correction term that has something to do with the boundary condition on S. As we would expect, this correction term goes to zero as the surface S goes to infinity, which recovers the basic solution for a source in empty, infinite space.

As it stands, this expression for Φ is quite general in the sense that nothing has yet been said about the type of boundary condition appearing in the surface term. Let us now examine two particular cases: (1) the potential Φ itself is specified on the boundary. This type of situation is known as a Dirichlet problem. (2) Instead of the potential, the electric field, or the normal derivative of Φ, is known on S. This constitutes a Neumann boundary condition. As it turns out, either one of these situations results in a *unique* solution.

To see how this comes about, assume for the time being that there exist two solutions Φ_1 and Φ_2 satisfying the *same* boundary condition. Then, inside the volume of solution, we must have

$$\vec{\nabla}^2\Phi_1 = \vec{\nabla}^2\Phi_2 = -4\pi\rho\,, \tag{2.28}$$

so that

$$\vec{\nabla}^2(\Phi_1 - \Phi_2) = 0\,. \tag{2.29}$$

On the boundary S, either $\Phi_1 - \Phi_2 = 0$ or $\partial(\Phi_1 - \Phi_2)/\partial n = 0$, depending on which type of problem—Dirichlet or Neumann—we are faced with. Relating the solution for Φ to its prescribed value on the surface is most easily accomplished with the use of Green's first identity, which was derived specifically for this purpose. Although the form of Equation (2.24) was cast with the intention of connecting Φ to ψ, it is in fact quite general and is valid for any two scalar functions χ_1 and χ_2. And so putting $\chi_1 = \chi_2 = \Phi_1 - \Phi_2$ in this equation, we get

$$\int_V \left[(\Phi_1 - \Phi_2)\vec{\nabla}'^2(\Phi_1 - \Phi_2) + |\vec{\nabla}(\Phi_1 - \Phi_2)|^2 \right] d^3x' =$$
$$\oint_S (\Phi_1 - \Phi_2)\frac{\partial}{\partial n}(\Phi_1 - \Phi_2)\, da'\,. \tag{2.30}$$

Evidently, either type of boundary condition results in the following constraint:

$$\int_V |\vec{\nabla}'(\Phi_1 - \Phi_2)|^2\, d^3x' = 0\,. \tag{2.31}$$

Since $|\vec{\nabla}'(\Phi_1-\Phi_2)|^2 \geq 0$, this equation necessarily implies that $\vec{\nabla}'(\Phi_1-\Phi_2) = 0$, which in turn has the solution $\Phi_1-\Phi_2 = $ constant everywhere within the domain of validity. For Dirichlet problems, $\Phi_1 - \Phi_2 = 0$ on S, so that the constant must be zero and $\Phi_1 = \Phi_2$ everywhere. For Neumann conditions, Φ_1 and Φ_2 are identical apart from an arbitrary additive constant that does not affect $\mathbf{E}$.

With Equation (2.25), the proof of a unique solution, and our understanding of Equation (2.27), we now have the necessary tools to begin considering formal solutions of electrostatic boundary-value problems using Green functions.[9] Although providing us with much needed insight, Equation (2.27) by itself is quite impractical for this purpose, since it yields a solution only when both Φ and $\partial\Phi/\partial n$ are known on S. But specifying these quantities a priori over-constrains the problem, since we have seen that either a Dirichlet or a Neumann condition leads to a unique solution. This doesn't mean, of course, that both Φ and $\partial\Phi/\partial n$ cannot be known simultaneously; on the contrary, a unique, well-defined solution has both. It is just that we cannot set the conditions on Φ and $\partial\Phi/\partial n$ independently of each other without knowing the solution beforehand. This is why we must modify our approach somewhat, so that the final equation we derive from (2.25) has a surface term that involves either Φ or $\partial\Phi/\partial n$, but not both.

In deriving Equation (2.27), we made use of the fact that $\vec{\nabla}'^2(|\mathbf{x} - \mathbf{x}'|^{-1}) = -4\pi\delta^3(\mathbf{x} - \mathbf{x}')$ in order to extract Φ from the general integral. This is the way to solve Poisson's equation when the boundary conditions lie at infinity so that the "remainder" terms in Green's theorem are irrelevant. With meaningful boundary conditions, we expect the potential to be modified, and one way to describe this mathematically is to put

$$\Phi \propto \frac{1}{|\mathbf{x} - \mathbf{x}'|} + F(\mathbf{x}, \mathbf{x}') \,, \tag{2.32}$$

as we alluded to previously. $F(\mathbf{x}, \mathbf{x}')$ is a function that must satisfy Poisson's equation and the boundary condition. Since ψ was originally introduced to represent the potential per unit charge, it makes sense here to now choose

$$\psi(\mathbf{x}, \mathbf{x}') = \frac{1}{|\mathbf{x} - \mathbf{x}'|} + F(\mathbf{x}, \mathbf{x}') \,. \tag{2.33}$$

Written in this way, ψ becomes the general Green function $G(\mathbf{x}, \mathbf{x}')$, and forcing it to satisfy the condition

$$\vec{\nabla}'^2 G(\mathbf{x}, \mathbf{x}') = -4\pi\,\delta^3(\mathbf{x} - \mathbf{x}') \tag{2.34}$$

(i.e., requiring it to represent the potential for a unit point charge), preserves many of the benefits of our earlier use of Equation (2.25) to derive (2.27). We note here that $F(\mathbf{x}, \mathbf{x}')$ must therefore satisfy Laplace's equation

$$\vec{\nabla}'^2 F(\mathbf{x}, \mathbf{x}') = 0 \tag{2.35}$$

inside the volume of interest, which is fully consistent with the fact that all the imaginary charges, which give rise to $F(\mathbf{x}, \mathbf{x}')$, must lie *outside* the boundary

---

[9] A good general discussion of the Green function technique may be found in Morse and Feshbach (1953), and Butkov (1968).

surface S. The presence of this function provides us with the flexibility to eliminate one or the other of the surface integrals in Green's theorem and to thereby obtain an expression for Φ that involves only Dirichlet or Neumann boundary conditions.

When we substitute $G(\mathbf{x}, \mathbf{x}')$ for ψ in Equation (2.25), the resulting expression for Φ is

$$\Phi(\mathbf{x}) \;=\; \int_V \rho(\mathbf{x}')\, G(\mathbf{x}, \mathbf{x}')\, d^3 x'$$

$$+\frac{1}{4\pi} \oint_S \left[G(\mathbf{x}, \mathbf{x}') \frac{\partial \Phi}{\partial n} - \Phi(\mathbf{x}') \frac{\partial G(\mathbf{x}, \mathbf{x}')}{\partial n} \right] da' . \qquad (2.36)$$

For Dirichlet boundary conditions, we know Φ on S, and so we require that $G(\mathbf{x}, \mathbf{x}') \to G_D(\mathbf{x}, \mathbf{x}')$, where

$$G_D(\mathbf{x}, \mathbf{x}') = 0 \qquad (2.37)$$

for $\mathbf{x}'$ on S. In that case,

$$\Phi(\mathbf{x}) = \int_V \rho(\mathbf{x}') G_D(\mathbf{x}, \mathbf{x}')\, d^3 x' - \frac{1}{4\pi} \oint_S \Phi(\mathbf{x}') \frac{\partial G_D}{\partial n}\, da' . \qquad (2.38)$$

For Neumann boundary conditions, we can use a similar procedure, but with one minor difference. We cannot use $\partial G/\partial n = 0$ on S because this is inconsistent with $\vec{\nabla}'^2 G(\mathbf{x}, \mathbf{x}') = -4\pi\, \delta^3(\mathbf{x} - \mathbf{x}')$. In other words, the outflux of G cannot be zero when there is a source enclosed by S. To see this rigorously, we return momentarily to the Divergence theorem, which we use as follows:

$$\int_V \vec{\nabla}'^2 G(\mathbf{x}, \mathbf{x}')\, d^3 x' \;=\; -4\pi = \int_V \vec{\nabla}' \cdot [\vec{\nabla}' G(\mathbf{x}, \mathbf{x}')]\, d^3 x'$$

$$= \oint_S \vec{\nabla}' G(\mathbf{x}, \mathbf{x}') \cdot \hat{n}\, da'$$

$$= \oint_S \frac{\partial G(\mathbf{x}, \mathbf{x}')}{\partial n}\, da' . \qquad (2.39)$$

Therefore,

$$\oint_S \frac{\partial G(\mathbf{x}, \mathbf{x}')}{\partial n}\, da' = -4\pi \neq 0 , \qquad (2.40)$$

and the simplest boundary condition we can use is thus

$$\frac{\partial G_N(\mathbf{x}, \mathbf{x}')}{\partial n} = -\frac{4\pi}{S} \qquad (2.41)$$

for $\mathbf{x}'$ on S. S is the total surface area, so the right-hand side represents a "weighting" over the entire boundary. Then,

$$\Phi(\mathbf{x}) = \int_V \rho(\mathbf{x}')\, G_N(\mathbf{x}, \mathbf{x}')\, d^3x' + \frac{1}{4\pi} \oint_S \frac{\partial \Phi}{\partial n}\, G_N\, da' + \langle \Phi \rangle_S \,, \qquad (2.42)$$

where

$$\langle \Phi \rangle_S \equiv \frac{1}{S} \oint_S \Phi(\mathbf{x}')\, da' \qquad (2.43)$$

is the average value of the potential over the whole surface S. In most cases, S is extremely large (or even infinite), in which case $\langle \Phi \rangle_S \to 0$.

In summarizing the results of this section, we reiterate the role of the key term we have added to describe the Coulomb potential from a charge inside a closed surface S. $F(\mathbf{x}, \mathbf{x}')$ is a solution of the Laplace equation within the volume bounded by S. It therefore represents the potential of a system of charges *external* to V. For every parcel of charge $\rho(\mathbf{x}')\, d^3x'$ at $\mathbf{x}'$, there exists an external distribution of charge that, combined with $\rho(\mathbf{x}')\, d^3x'$, satisfies the homogeneous boundary condition (i.e., $\Phi = 0$) on S for Dirichlet problems, or yields a simple average value of the normal derivative of the surface potential for Neumann conditions. The determination of $F(\mathbf{x}, \mathbf{x}')$ is thus equivalent to the method of images discussed in the previous section, but it is obviously much more powerful for complicated charge distributions $\rho(\mathbf{x}')$. It is extremely important to understand that no matter how the source is distributed, finding the Green function is completely independent of $\rho(\mathbf{x}')$. $G(\mathbf{x}, \mathbf{x}')$ depends exclusively on the geometry of the problem, because its function is that of a "template" potential, not the actual potential for a given physical setup. In other words, $G(\mathbf{x}, \mathbf{x}')$ is the potential due to a unit charge, positioned arbitrarily within the volume of solution, consistent with either $G_D = 0$ or $\partial G_N / \partial n = -4\pi/S$ on the surface. The potential is then the convolution of this template with the given $\rho(\mathbf{x}')$. When either $\Phi \neq 0$ or $\partial \Phi / \partial n \neq 0$ on the boundary, the correction to Φ is applied by the surface term, whose form depends on whether we are dealing with Dirichlet or Neumann conditions. Unfortunately, this aspect of the Green function is often overlooked, or forgotten, but it is the most important ingredient of this method of solution. There are very few symmetries for which the Green function is easily derivable—we will look at two of them in the examples that follow—yet there exists an abundance of problems for which this technique works, mainly because we are free to choose a wide variety of charge densities and boundary conditions. A summary of Dirichlet and Neumann problems follows.

Dirichlet Problems

$$\Phi(\mathbf{x}) = \int_V \rho(\mathbf{x}')\, G_D(\mathbf{x},\mathbf{x}')\, d^3x' - \frac{1}{4\pi} \oint_S \Phi(\mathbf{x}')\, \frac{\partial G_D}{\partial n}\, da'$$

with Φ specified on S (Equation [2.38]), and

$$G_D(\mathbf{x},\mathbf{x}') = \frac{1}{|\mathbf{x}-\mathbf{x}'|} + F(\mathbf{x},\mathbf{x}')\,, \tag{2.44}$$

where (from Equation [2.35])

$$\vec{\nabla}'^2 F(\mathbf{x},\mathbf{x}') = 0\,,$$

and (from Equation [2.37])

$$G_D(\mathbf{x},\mathbf{x}') = 0 \quad \text{for } \mathbf{x}' \text{ on } S\,.$$

Neumann Problems

$$\Phi(\mathbf{x}) = \int_V \rho(\mathbf{x}')\, G_D(\mathbf{x},\mathbf{x}')\, d^3x' + \frac{1}{4\pi} \oint_S \frac{\partial \Phi}{\partial n}\, G_N(\mathbf{x},\mathbf{x}')\, da' + \langle\Phi\rangle_S$$

with $\partial\Phi/\partial n$ given on S (Equation [2.42]), and

$$G_N(\mathbf{x},\mathbf{x}') = \frac{1}{|\mathbf{x}-\mathbf{x}'|} + F(\mathbf{x},\mathbf{x}')\,, \tag{2.45}$$

where (again from Equation [2.35])

$$\vec{\nabla}'^2 F(\mathbf{x},\mathbf{x}') = 0\,,$$

and (from Equation [2.41])

$$\frac{\partial G_N(\mathbf{x},\mathbf{x}')}{\partial n} = -\frac{4\pi}{S} \quad \text{for } \mathbf{x}' \text{ on } S\,.$$

Example 2.1. Consider the volume in the half space $x \geq 0$, with $\Phi = \Phi_0$ on the surface S defined by $x = 0$. Suppose there is a charge q at $(a,0,0)$. Find the potential at any point $\vec{r} = \sqrt{x^2+y^2+z^2}\ \hat{r}$ in the domain of solution (Figure 2.3).

Here, we need $G_D = 0$ on S. Thus,

$$\left\{ \frac{1}{|\mathbf{x} - \mathbf{x'}|} + F(\mathbf{x}, \mathbf{x'}) \right\}_S = 0 , \tag{2.46}$$

or

$$F(\mathbf{x}, \mathbf{x'})|_S = -\left\{ \frac{1}{[(x - x')^2 + (y - y')^2 + (z - z')^2]^{1/2}} \right\}_S . \tag{2.47}$$

On S, $x' = 0$, so

$$F(\mathbf{x}, \mathbf{x'})|_S = \frac{-1}{[x^2 + (y - y')^2 + (z - z')^2]^{1/2}} . \tag{2.48}$$

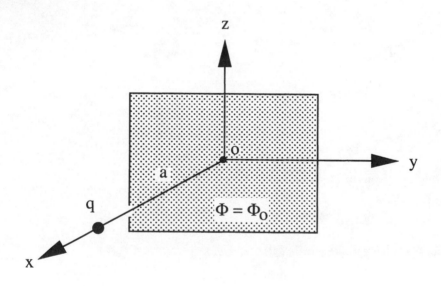

Figure 2.3 A charge q near a boundary surface maintained at a fixed potential Φ_0 on the yz-plane. The surface is assumed to be an infinite plane a distance a from q.

Can we now find an $F(\mathbf{x}, \mathbf{x'})$ that reduces to the form in Equation (2.48) at the surface? The planar symmetry and the analogy with the method of images would suggest that we try something like

$$F(\mathbf{x}, \mathbf{x'}) = -\left\{ \frac{1}{[(x \pm x')^2 + (y - y')^2 + (z - z')^2]^{1/2}} \right\}_S , \tag{2.49}$$

but which of $\pm x'$ do we choose? Physically, the two signs mean different things: a $-$ would indicate that we are placing the imaginary charge at $x' = x$, whereas a $+$ would imply that the imaginary charge is positioned at $x' = -x$. But remember that we absolutely cannot put any fictitious charges within the volume of solution, so that only the second choice is a valid possibility. Thus,

$$F(\mathbf{x}, \mathbf{x}') = \frac{-1}{[(x + x')^2 + (y - y')^2 + (z - z')^2]^{1/2}} \ . \tag{2.50}$$

We are still not done with this, because we must confirm that our choice of F satisfies Laplace's equation within the domain of solution. In this case, showing that $\vec{\nabla}'^2 F(\mathbf{x}, \mathbf{x}') = 0$ is trivial, and so the Dirichlet-Green function for this geometry (we emphasize that it is for this *geometry*, not for this given charge distribution) is

$$
\begin{aligned}
G_D(\mathbf{x}, \mathbf{x}') \ = \ & [(x - x')^2 + (y - y')^2 + (z - z')^2]^{-1/2} \\
& -[(x + x')^2 + (y - y')^2 + (z - z')^2]^{-1/2} \ . \tag{2.51}
\end{aligned}
$$

Since $\rho(\mathbf{x}') = q\,\delta^3(\mathbf{x}' - [a, 0, 0])$ (the units are consistent since the three-dimensional delta function has units of [length]$^{-3}$), this problem is solved with the following expression for the potential:

$$\Phi(\mathbf{x}) = \int_V q\,\delta^3(\mathbf{x} - [a, 0, 0])\, G_D(\mathbf{x}, \mathbf{x}')\, d^3x' - \frac{1}{4\pi} \oint_S \Phi_0 \frac{\partial G_D}{\partial n}\, da' \ . \tag{2.52}$$

From here the steps are routine, and we do not need to show them in detail. The single remaining possible pitfall is the evaluation of $\partial G_D / \partial n$, which requires a careful consideration of what the outward normal unit vector $\hat{n}$ is. In this problem, $\hat{n}$ points in the $-x'$ direction (not $+x'$!), so that

$$\left.\frac{\partial G_D}{\partial n}\right|_S = -\left.\frac{\partial G_D}{\partial x'}\right|_{x'=0} = \frac{-2x}{[x^2 + (y - y')^2 + (z - z')^2]^{3/2}} \ . \tag{2.53}$$

The final solution is therefore

$$
\begin{aligned}
\Phi(\mathbf{x}) \ = \ & q\left\{ [(x - a)^2 + y^2 + z^2]^{-1/2} - [(x + a)^2 + y^2 + z^2]^{-1/2} \right\} \\
& +\frac{\Phi_0 x}{2\pi} \int_{-\infty}^{\infty} \int_{-\infty}^{\infty} \frac{dy'\, dz'}{[x^2 + (y - y')^2 + (z - z')^2]^{3/2}} \ . \tag{2.54}
\end{aligned}
$$

As a final useful exercise, one should use this expression to determine the charge density σ on the surface, as discussed in the section on the method of images.

Example 2.2. To close this section, we will solve a Dirichlet problem for the interior region of a sphere. Imagine that the surface of the sphere (radius a)

is maintained at a potential $\Phi(a, \theta, \phi)$, but that there are no charges inside or outside. The fact that the given charge distribution is zero here can sometimes confuse the issue of how the Green function is derived and how it is used. After all, if $\rho(\mathbf{x}') = 0$, doesn't this mean that the potential due to the source is zero? Yes, but that is not the point! We must remember that the Green function is entirely geometric, and acts as a template whose properties are independent of ρ. If $\rho(\mathbf{x}') = 0$, this simply means that the contribution to $\Phi(\mathbf{x})$ from the source is zero, but the effects of the boundary, as given by the surface term in the general expression for Φ, cannot be neglected. The Green function is pivotal to the determination of both contributions.

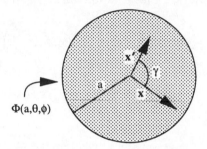

Figure 2.4 Definition of the coordinate system used to solve the interior Dirichlet problem for a sphere of radius a. Although there are no charges, the spherical surface is maintained at a potential $\Phi(a, \theta, \phi)$.

Suppose we put a unit charge at $\mathbf{x}'$ within the sphere (Figure 2.4). By symmetry, we expect that the image charge invoked to cancel the effect of this unit charge on the surface (i.e., to give us the required boundary condition) should lie on a ray from the origin passing through $\mathbf{x}'$. For the interior problem, the image charge q_0 must lie outside the sphere. Let this image charge be positioned at $\mathbf{x}_0'$. Then, the Green function is

$$G_D(\mathbf{x}, \mathbf{x}') = \frac{1}{|\mathbf{x} - \mathbf{x}'|} + \frac{q_0}{|\mathbf{x} - \mathbf{x}_0'|} . \tag{2.55}$$

We must now choose $\mathbf{x}_0'$ and q_0 so that $G_D = 0$ on the surface. Let $\hat{n}$ be a unit vector in the direction $\mathbf{x}$ (the point of observation) and $\hat{n}'$ in the direction $\mathbf{x}'$. It follows that

$$G_D(\mathbf{x}, \mathbf{x}') = \frac{1}{|r\hat{n} - r'\hat{n}'|} + \frac{q_0}{|r\hat{n} - r_0'\hat{n}'|} , \tag{2.56}$$

where $\mathbf{x} = r\hat{n}$, $\mathbf{x}' = r'\hat{n}'$, and $\mathbf{x}_0' = r_0'\hat{n}'$. Let's factor out the r and r_0', so that

$$G_D(\mathbf{x}, \mathbf{x}') = \frac{1}{r|\hat{n} - (r'/r)\hat{n}'|} + \frac{q_0}{r_0'|\hat{n}' - (r/r_0')\hat{n}|} , \qquad (2.57)$$

which on the surface reduces to

$$G_D(\mathbf{x}, \mathbf{x}')|_S = \frac{1}{a|\hat{n} - (r'/a)\hat{n}'|} + \frac{q_0}{r_0'|\hat{n}' - (a/r_0')\hat{n}|} . \qquad (2.58)$$

It's clear that if we choose $\mathbf{x}_0'$ and q_0 such that

$$\frac{q_0}{r_0'} = -\frac{1}{a} \qquad (2.59)$$

and

$$\frac{a}{r_0'} = \frac{r'}{a} , \qquad (2.60)$$

then $G_D = 0$ on S for all values of $\hat{n} \cdot \hat{n}'$. Thus, the magnitude and position of the image charge should be

$$q_0 = -\frac{a}{r'} \qquad (2.61)$$

and

$$r_0' = \frac{a^2}{r'} , \qquad (2.62)$$

respectively. Expressing the positions using spherical coordinates, we therefore get

$$G_D(\mathbf{x}, \mathbf{x}') = [r^2 - r'^2 - 2rr' \cos\gamma]^{-1/2} - [r^2 r'^2/a^2 + a^2 - 2rr' \cos\gamma]^{-1/2} , \quad (2.63)$$

where $\cos\gamma \equiv \cos\theta \cos\theta' + \sin\theta \sin\theta' \cos(\phi - \phi')$. Noting that $\hat{n}'$ is the outward unit normal, which here points in the $+\hat{r}'$ direction, it is straightforward to show that

$$\left.\frac{\partial G_D}{\partial n'}\right|_S = \left.\frac{\partial G_D}{\partial r'}\right|_{r'=a} = \frac{(r^2 - a^2)}{a(r^2 + a^2 - 2ar \cos\gamma)^{3/2}} . \qquad (2.64)$$

And so the solution to the Laplace equation *inside* the sphere is

$$\Phi(\mathbf{x}) = -\frac{1}{4\pi} \oint_S \Phi(a, \theta', \phi') \frac{a(r^2 - a^2)}{(r^2 + a^2 - 2ar \cos\gamma)^{3/2}} d\Omega' , \qquad (2.65)$$

where $da' = a^2 d\Omega'$ and $d\Omega'$ is the element of solid angle at (a, θ', ϕ').

2.1.3   Expansions with Orthonormal Functions

In this method, the potential and other variables are written as series of orthonormal functions, chosen to match the inherent symmetry. The properties of these functions allow us to calculate the expansion coefficients trivially and we can get a solution to arbitrary accuracy, provided the set of orthonormal functions is complete (see below). The functions, U_n, must be *orthonormal* on a prescribed interval (a, b), meaning that

$$\int_a^b U_n^*(\xi)U_m(\xi)\, d\xi = \delta_{nm}\,, \tag{2.66}$$

for any two members, U_n, U_m, of the set. The set must also be *complete*, in the sense that any arbitrary function $f(\xi)$ can be expanded as a series of the U_n:

$$f(\xi) = \sum_{n=1}^{N} a_n U_n(\xi)\,, \quad \text{where} \quad a_n = \int_a^b U_n^*(\xi)f(\xi)\, d\xi\,, \tag{2.67}$$

there being a finite number N_{max} such that for $N > N_{max}$, the mean square error

$$M_N \equiv \int_a^b \left| f(\xi) - \sum_{n=1}^{N} a_n U_n(\xi) \right|^2 d\xi \tag{2.68}$$

can be made smaller than any arbitrarily small positive quantity.

This technique is useful when the problem has an obvious symmetry so that an appropriate coordinate system and a set of functions can be matched to it. The most direct approach in solving Laplace's equation is to separate the variables, which has been accomplished in 11 different coordinate systems for the three-dimensional Laplacian operator. We list in Table 2.1 for reference some of the most commonly used ones. The procedure for separating the variables is straightforward and has been described in many other contexts; it does not need to be repeated in detail here. Instead, we will focus on the physics and general form of the solutions.

Cartesian Coordinates

In Cartesian coordinates, the potential is given by the expansion

$$\Phi(x, y, z) = \sum_{n,m} a_{nm} \exp(\pm i\alpha_n x) \exp(\pm i\beta_m y) \exp(\pm \sqrt{\alpha_n^2 + \beta_m^2}\, z)\,. \tag{2.69}$$

Symmetry	Coordinate System	Orthonormal Set
Cartesian	Cartesian (x, y, z)	Fourier series $\exp(\pm i\mathbf{k} \cdot \mathbf{x})$
Spherical	Polar (r, θ, ϕ)	Spherical harmonics, $Y_{lm}(\theta, \phi)$, which include the Legendre polynomials, $P_n(\cos\theta)$
Cylindrical	Cylindrical (η, ϕ, z)	Bessel functions, J_ν
Elliptical	Confocal elliptical	Mathieu functions and Hermite functions

Table 2.1 The most commonly used coordinate systems for the three-dimensional Laplacian operator.

The choice of plus or minus in these arguments is dictated by the boundary conditions. For specific applications, one or both of the $\exp(\pm i\alpha_n x)$ and $\exp(\pm i\beta_m y)$ may be replaced by $\sin\alpha_n x$ or $\cos\beta_m y$.

Example 2.3. As a specific example, let us consider the interior Dirichlet problem for a rectangular boundary (Figure 2.5). Since there is no z-dependence here, the solution will contain only terms like

$$\exp(\pm i\alpha x) \qquad \text{and} \qquad \exp(\pm i\beta y)\,, \qquad (2.70)$$

and in addition, we must have $\sqrt{\alpha^2 + \beta^2} = 0$, which means that $\alpha = \pm i\beta$. Thus, either α or β may be imaginary, but not both. If α is real, the possible combinations for the exponential terms are

$$\cos\alpha x = \frac{1}{2}[\exp(i\alpha x) + \exp(-i\alpha x)]\,,$$

$$\sin\alpha x = \frac{1}{2}[\exp(i\alpha x) - \exp(-i\alpha x)]\,,$$

$$\cosh|\alpha|y = \frac{1}{2}[\exp(|\alpha|y) + \exp(-|\alpha|y)]\,,$$

$$\sinh|\alpha|y = \frac{1}{2}[\exp(|\alpha|y) - \exp(-|\alpha|y)]\,. \qquad (2.71)$$

The functions for x and y are reversed if instead α is imaginary and β is real. In principle, the solution to this problem can therefore be written as a series composed of terms like the following:

(i) $\quad (A \sin \lambda x + B \cos \lambda x)(C \sinh \mu y + D \cosh \mu y)$,

(ii) $\quad (A \sin \lambda x + B \cos \lambda x)(C \sin \mu y + D \cos \mu y)$,

(iii) $\quad (A \sinh \lambda x + B \cosh \lambda x)(C \sin \mu y + D \cos \mu y)$,

(iv) $\quad (A \sinh \lambda x + B \cosh \lambda x)(C \sinh \mu y + D \cosh \mu y)$.

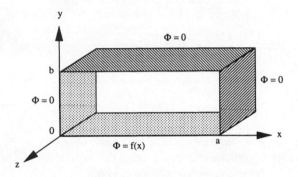

Figure 2.5 The rectangular region defining the domain of solution. The boundary conditions are independent of z and are specified as functions of x and y. The potential Φ is everywhere zero except on the xz-plane, where $\Phi = f(x)$.

But now let's look at the boundary conditions to see how many of these we can eliminate before we even begin to write down the expansion for Φ. In terms (iii) and (iv), the condition $\Phi(0, y) = 0$ requires $B = 0$. But then $\Phi(a, y) = 0$ also forces A to be zero, and so we can rule these out immediately. In terms (i) and (ii), the boundary value $\Phi(0, y) = 0$ requires $B = 0$. Since, moreover, $\Phi(a, y) = 0$, the sin function must have a node at $x = a$, meaning that

$$\lambda = \frac{n\pi}{a} \equiv \lambda_n . \tag{2.72}$$

Further, we can eliminate term (ii) using the requirement that $\alpha = i\beta$, which forces the y-functions to be hyperbolic if the x-functions are trigonometric, and in addition, forces $\mu = \lambda$. This is consistent with the fact that the boundary conditions in the y-direction are not periodic. So now the only possible expansion terms are

$$(\text{v}) \qquad A \sin\left(\frac{n\pi}{a} x\right) \left\{ C \sinh\left(\frac{n\pi}{a} y\right) + D \cosh\left(\frac{n\pi}{a} y\right) \right\} .$$

At $y = b$, we must have $C \sinh(n\pi b/a) + D \cosh(n\pi b/a) = 0$, which eliminates D in terms of C:

$$\sin\left(\frac{n\pi}{a} x\right) \left\{ \sinh\left(\frac{n\pi}{a} y\right) - \frac{\sinh(n\pi b/a)}{\cosh(n\pi b/a)} \cosh\left(\frac{n\pi}{a} y\right) \right\}$$

$$\sim \sin\left(\frac{n\pi}{a} x\right) \sinh\left\{ \frac{n\pi}{a} (b - y) \right\} .$$

This satisfies all the boundary conditions, except at $y = 0$, which we must use to evaluate the coefficients in the series expansion. As a result of this preliminary matching of possible terms with the given constraints, we have reduced the general expression for Φ (see Equation [2.69]) to the following:

$$\Phi(x, y) = \sum_{n=1}^{\infty} F_n \sin\left(\frac{n\pi}{a} x\right) \sinh\left\{ \frac{n\pi}{a} (b - y) \right\} . \tag{2.73}$$

Finally, at $y = 0$, we have $\Phi(x, 0) = f(x)$, so that

$$f(x) = \sum_{n=1}^{\infty} F_n \sin\left(\frac{n\pi}{a} x\right) \sinh\left(\frac{n\pi}{a} b\right) , \tag{2.74}$$

and by orthogonality of the set of functions $\sin(n\pi x/a)$, the problem is solved with the evaluation of the coefficients F_n according to the expression

$$F_n \sinh\left(\frac{n\pi}{a} b\right) = \frac{2}{a} \int_0^a f(x) \sin\left(\frac{n\pi}{a} x\right) dx . \tag{2.75}$$

Cylindrical Coordinates

In cylindrical coordinates, the Laplace equation is

$$\vec{\nabla}^2 \Phi = \frac{1}{\eta} \frac{\partial}{\partial \eta} \left(\eta \frac{\partial \Phi}{\partial \eta} \right) + \frac{1}{\eta^2} \frac{\partial^2 \Phi}{\partial \phi^2} + \frac{\partial^2 \Phi}{\partial z^2} = 0 . \tag{2.76}$$

Assuming that the dependences of Φ on the coordinates are separable, so that

$$\Phi(\eta, \phi, z) = R(\eta) F(\phi) Z(z) , \tag{2.77}$$

we obtain from this three separated ordinary differential equations:

$$\frac{d^2 F}{d\phi^2} + \nu^2 F = 0 \,, \tag{2.78}$$

$$\frac{d^2 Z}{dz^2} - k^2 Z = 0 \,, \tag{2.79}$$

$$\frac{d^2 R}{d\eta^2} + \frac{1}{\eta}\frac{dR}{d\eta} + \left(k^2 - \frac{\nu^2}{\eta^2}\right) R = 0 \,, \tag{2.80}$$

where ν and k are the separation constants. The solutions to these are straightforward, but several key points merit our additional attention. To begin with, the solution to Equation (2.78) is

$$F(\phi) = \exp(\pm i\nu\phi) \,. \tag{2.81}$$

If we insist that F should be a single-valued function on the full range of ϕ, then ν is restricted to being an integer. That is, we want $F(\phi + 2\pi) = F(\phi)$, which implies that $\exp(\pm i\nu 2\pi) = 1$ and therefore $\nu = 0, \pm 1, \pm 2, \ldots$. This is but one example of the eigenvalues that result from the solution to Laplace's equation subject to certain boundary conditions. Many of us are familiar with a similar situation in quantum mechanical systems, where the boundary conditions impose specific eigenvalue constraints on the wavefunction, related to the angular momentum.

One solution to the radial equation (2.80) is a Bessel function of order ν:

$$J_\nu(k\eta) = \left(\frac{k\eta}{2}\right)^\nu \sum_{j=0}^\infty \frac{(-1)^j}{j!\,\Gamma(j+\nu+1)}\left(\frac{k\eta}{2}\right)^{2j} \,, \tag{2.82}$$

where $\Gamma(u)$ is the gamma function of u. Of course, Equation (2.80) is a second-order ODE, so it has a second linearly independent solution, which is often taken to be the Neumann function

$$N_\nu(k\eta) = \frac{J_\nu(k\eta)\cos(\nu\pi) - J_{-\nu}(k\eta)}{\sin\nu\pi} \,. \tag{2.83}$$

Here, $J_{-\nu}(k\eta) = (-1)^\nu J_\nu(k\eta)$, a definition that is valid only when ν is an integer. In this case, however, $N_\nu(k\eta)$ becomes indeterminate. Using l'Hospital's rule for indeterminate forms and a power series expansion for $J_\nu(k\eta)$, one can show that $N_\nu(k\eta)$ exhibits a logarithmic functionality that clearly makes it independent of $J_\nu(k\eta)$. The point to emphasize here is that whereas J_ν is regular at the origin, N_ν is singular. This is crucial in deciding which set of functions to use in the expansion for Φ subject to a given set of boundary conditions. For example, the interior region of a cylinder could not be described using Neumann functions since these would diverge as $\eta \to 0$.

With these caveats in mind, the general solution to Laplace's equation in a cylindrical geometry is

$$\Phi(\eta, \phi, z) = \sum_{\nu} \sum_{n=1}^{\infty} \{a_{\nu n} J_\nu(k_{\nu n} \eta) + b_{\nu n} N_\nu(k_{\nu n} \eta)\}$$

$$\times \{c_{\nu n} \sinh(k_{\nu n} z) + d_{\nu n} \cosh(k_{\nu n} z)\} \exp(i\nu\phi) , \quad (2.84)$$

where $k_{\nu n}\eta$ is the nth root of J_ν at the radial boundary of the problem. For example, if we are treating the interior solution for a cylinder of radius a, then $k_{\nu n}$ is the nth root of J_ν divided by a. We shall elaborate on this during our discussion of the following problem. The constants $k_{\nu n}$ can sometimes be imaginary. In this case, the J_ν and N_ν are the modified (or hyperbolic) Bessel functions, and the z-dependence is given in terms of sin and cosine.

Example 2.4. The application we will consider here is the interior Dirichlet problem for a cylinder, in which the surface potential is specified to be zero on the bottom face and the side of the cylinder and to have the functional form $f(\eta)$ on the top face (Figure 2.6). The cylinder is assumed to have a radius a and length L, and to be oriented such that the z-axis is parallel to its axis of symmetry. The boundary condition is cylindrically symmetric, so we expect Φ to be independent of the azimuthal angle ϕ. That immediately restricts ν to be zero. In addition, we know that

$$N_0(k\eta) \to \frac{2}{\pi} \left\{ \ln\left(\frac{k\eta}{2}\right) + 0.5772... \right\} \qquad \text{for } k\eta \ll 1 , \quad (2.85)$$

so that Φ can remain finite at $\eta = 0$ only if $b_{\nu n} = 0$ for all values of n. Since $\Phi = 0$ at $z = 0$, $d_{\nu n}$ must be zero as well, and the solution is therefore effectively reduced to

$$\Phi(\eta, \phi, z) = \sum_{n=1}^{\infty} a_n J_0(k_n \eta) \sinh(k_n z) . \quad (2.86)$$

Like the Legendre polynomials, the Bessel functions form a complete set over a finite interval. The Fourier-Bessel theorem[10] states that a function $g(\eta)$ can

---

[10]Churchill (1963) gives an extensive discussion of the mathematical theory of Fourier series and integrals, and expansions with orthonormal functions.

be represented as an infinite series of Bessel functions over the interval $(0, a)$ according to

$$g(\eta) = \sum_{n=1}^{\infty} a_{\nu n} J_\nu \left(x_{\nu n} \frac{\eta}{a} \right) , \qquad (2.87)$$

where $x_{\nu n}$ is the nth root of $J_\nu(x)$, that is,

$$J_\nu(x_{\nu n}) = 0 . \qquad (2.88)$$

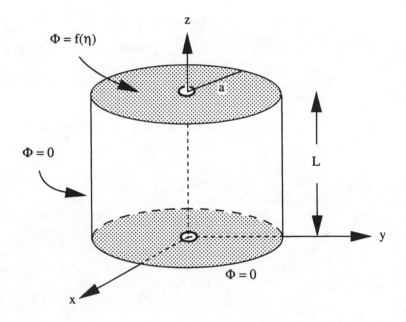

Figure 2.6 Interior Dirichlet problem for a cylinder of finite length L and radius a aligned with the z-axis. The sides and bottom face are maintained at zero potential, whereas the top face has an arbitrary, prescribed potential $f(\eta)$.

This is analogous to a trigonometric series, where we would use the expansion

$$g(x) = \sum_{n=1}^{\infty} A_n \sin \left(\frac{n\pi}{a} x \right) \qquad (2.89)$$

for a function $g(x)$ on the interval $(0, a)$. Here, the nth root of $\sin(x)$ is $n\pi$.

The final step is to evaluate the expansion coefficients in Equation (2.86), which we determine by using the Bessel function orthogonality condition

$$\int_0^a \eta J_\nu \left(x_{\nu m} \frac{\eta}{a} \right) J_\nu \left(x_{\nu n} \frac{\eta}{a} \right) d\eta = \left(\frac{a^2}{2} \right) [J_{\nu+1}(x_{\nu n})]^2 \, \delta_{mn} \, . \tag{2.90}$$

At $z = L$, the potential is just $f(\eta)$, so that

$$\Phi(\eta, \phi, L) = f(\eta) = \sum_{n=1}^{\infty} a_n J_0 \left(x_{0n} \frac{\eta}{a} \right) \sinh \left(\frac{x_{0n} L}{a} \right) \, . \tag{2.91}$$

Thus, multiplying both sides by $\eta \, J_0(x_{0n}\eta/a)$ and integrating over η from 0 to a, we get

$$a_n = \frac{1}{\sinh(x_{0n}L/a)} \frac{2}{a^2 J_1^2(x_{0n})} \int_0^a \eta \, f(\eta) \, J_0 \left(x_{0n} \frac{\eta}{a} \right) d\eta \, , \tag{2.92}$$

and this in principle solves the problem.

Spherical Coordinates

The procedure for this geometry is strictly analogous to that of the previous case. The separation of variables produces a general expansion for Φ that involves power-law terms for the radial dependence, spherical harmonics for the angular functionality and a sinusoidal variation in the azimuthal direction:

$$\Phi(\mathbf{r}, \theta, \phi) = \frac{1}{r} \sum_{l,m} \left(a_{lm} \, r^{l+1} + b_{lm} \, r^{-l} \right) Y_{lm}(\theta, \phi) \, , \tag{2.93}$$

where

$$Y_{lm}(\theta, \phi) = \sqrt{\frac{2l+1}{4\pi} \frac{(l-m)!}{(l+m)!}} \, P_l^m(\cos\theta) \, \exp(im\phi) \, . \tag{2.94}$$

As was the case for the cylindrical Laplace equation, the requirement that the potential be single-valued forces m to be an integer. The θ equation yields a converging solution at $\cos\theta = \pm 1$ only if P_l^m are the associated Legendre functions, which have a finite number of terms (see, e.g., Magnus, Oberhettinger, and Soni 1966). In addition, this requirement of convergence at $\cos\theta = \pm 1$ forces m to be bounded by l, such that its permitted range is $-l, -(l-1), \ldots, 0, \ldots, (l-1), l$. When $m \neq 0$, the associated Legendre function can be evaluated from the general expression

$$P_l^m(x) = (-1)^m \left(1 - x^2 \right)^{m/2} \frac{d^m}{dx^m} \, P_l(x) \, , \tag{2.95}$$

where $P_l(x)$ is the solution to the equation

$$\frac{d}{dx}\left\{(1-x^2)\frac{dP}{dx}\right\} + l(l+1)P = 0 . \tag{2.96}$$

The first few such polynomials are

$$P_0(x) = 1 ,$$

$$P_1(x) = x ,$$

$$P_2(x) = (3x^2 - 1)/2 . \tag{2.97}$$

An illustration of how these expressions are used in problems with spherical symmetry is provided in Example 2.5 of § 2.2.2 below. With this we end our attention to the subject of electrostatics. However, before we move on to consider time-dependent electromagnetic phenomena, and then go on to develop the field theoretic structure of electrodynamics, we will study the equally important and intricate topic of magnetostatics. As it turns out, once we have identified the essential physics and have derived the necessary equations, we will learn that the "mechanics" of finding magnetostatic solutions is surprisingly similar to the procedures we have outlined above. Thus, the applicability of these methods extends well beyond the confines of the problems we have examined here.

2.2 MAGNETOSTATICS

The word "magnetostatics" sometimes confuses people, because it appears that the term "statics" implies an absence of motion and therefore currents. In reality, we deal with steady-state situations, which require only that the currents and other physical components are *time-independent*. Since free magnetic charges either do not exist or are extremely difficult to find, every-day magnetic behavior is closely connected to the nature of charge currents (i.e., moving charges). As we showed in deriving Equation (1.32), charge conservation requires that the time rate of change of charge density ρ be equal to the net outflux of current density per unit volume. With steady-state magnetic phenomena, the charge density must be constant, for which $\vec{\nabla} \cdot \mathbf{J} = 0$. This says that the streamlines of current flow do not have an origin or an end. In two-dimensional space, they must therefore extend to infinity in both directions, or they must close on themselves (see Figure 2.7). In three-dimensional space, the streamlines need not form closed loops to satisfy this condition, unless some special symmetry is imposed.

The general equations governing magnetostatics all follow from the Biot-Savart rule that relates the magnetic induction **B** to these currents. In differential form (cf. Equation [1.23]), the magnetic induction at the observation point **x** produced by an increment of current $I\,d\mathbf{l}'$ at $\mathbf{x}'$ is

$$d\mathbf{B} = \frac{I}{c}\frac{d\mathbf{l}' \times (\mathbf{x} - \mathbf{x}')}{|\mathbf{x} - \mathbf{x}'|^3}\,. \tag{2.98}$$

Since

$$\vec{\nabla}\left(\frac{1}{|\mathbf{x} - \mathbf{x}'|}\right) = -\frac{(\mathbf{x} - \mathbf{x}')}{|\mathbf{x} - \mathbf{x}'|^3}\,, \tag{2.99}$$

we get

$$\mathbf{B}(\mathbf{x}) = \frac{1}{c}\,\vec{\nabla}\times\int\frac{\mathbf{J}(\mathbf{x}')}{|\mathbf{x} - \mathbf{x}'|}\,d^3x' \tag{2.100}$$

(remembering that $\vec{\nabla}$ operates only on **x**, not $\mathbf{x}'$). It follows immediately that

$$\boxed{\vec{\nabla}\cdot\mathbf{B} = 0}\,. \tag{2.101}$$

We have seen this equation before (see Equation [1.14]), but we have derived it here in a different way. It should be stressed, however, that the underlying physics is identical. If we ignore the possible contribution to **B** from magnetic monopoles, the magnetic induction must be due to current loops and is therefore excluded from having points of origin and termination anywhere. Equation (2.101) is the first equation of magnetostatics and corresponds to the condition $\vec{\nabla}\times\mathbf{E} = 0$ in electrostatics. The second equation of magnetostatics follows from Maxwell's fourth equation with $\partial\mathbf{E}/\partial t = 0$:

$$\boxed{\vec{\nabla}\times\mathbf{B} = 4\pi\,\mathbf{J}/c}\,. \tag{2.102}$$

This also follows from Equation (2.100) with $\vec{\nabla}\cdot\mathbf{J} = 0$.

In matter, this equation is modified because the source function **J** has contributions from both free and bound charges. Following our discussion in § 1.2,

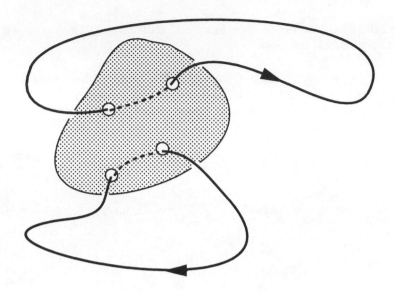

Figure 2.7 Closed loops of current threading a bounded region in two-dimensional space.

it is straightforward to see that Equation (2.102) should then be replaced by

$$\vec{\nabla} \times \mathbf{H} = \frac{4\pi}{c} \mathbf{J} \, , \qquad (2.103)$$

where the magnetic field $\mathbf{H}$ is defined in Equation (1.50) and reduces to the form given in (1.68) for linear, isotropic substances. We now have all the ingredients we will need in order to begin developing the techniques for solving magneto-statics problems. In the following subsections, we describe three different types of circumstances that commonly arise in steady-state situations.[11]

2.2.1   Method 1: The Magnetic Scalar Potential

In a region of space where $\mathbf{J} = 0$,

$$\vec{\nabla} \times \mathbf{H} = 0 \, . \qquad (2.104)$$

---

[11]A more extensive compilation of magnetostatics problems may be found in Smythe (1969).

Therefore, $\mathbf{H}$ must be the gradient of a scalar function Φ_m, which we shall call the magnetic scalar potential:

$$\mathbf{H} = -\vec{\nabla}\Phi_m . \tag{2.105}$$

If in addition the medium is linear and μ is constant, then with Equation (2.101), this becomes

$$\boxed{\vec{\nabla}^2\Phi_m = 0} . \tag{2.106}$$

The fact that this type of problem reduces once again to Laplace's equation means that, depending on the boundary conditions, we can use either or all of the techniques developed in § 2.1 for electrostatics. It should be remembered, however, that the physical meaning of Φ_m is different than that of Φ. Although it may not be immediately obvious, the integrand in Equation (1.23) looks very much like the expression for the solid angle subtended by the current loop at the observation point $\mathbf{x}$ (see Figure 2.8).

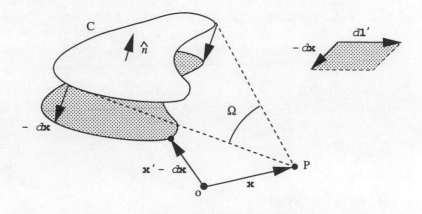

Figure 2.8 The solid angle Ω subtended at the observation point P by a closed current loop C, which is moved incrementally by $-d\mathbf{x}$ to generate a change in perspective.

The observer at $\mathbf{x}$ sees a solid angle

$$\Omega = \int d\Omega , \tag{2.107}$$

where

$$d\Omega = \frac{(da\,\hat{n}) \cdot (\mathbf{x}' - \mathbf{x})}{|\mathbf{x} - \mathbf{x}'|^3} \,, \tag{2.108}$$

and da is an element of enclosed surface area with unit normal $\hat{n}$. That is,

$$\Omega = -\int da\,\frac{\hat{n} \cdot (\mathbf{x} - \mathbf{x}')}{|\mathbf{x} - \mathbf{x}'|^3} \,. \tag{2.109}$$

Now imagine that the point $\mathbf{x}$ is moved by an increment $d\mathbf{x}$. How does Ω (and hence $\mathbf{B}$) change? This move is equivalent to moving the loop by an amount $-d\mathbf{x}$, as shown in Figure 2.8. The area seen at the observation point will thus change by a sum over the increments $-d\mathbf{x} \times d\mathbf{l}'$ around the loop. That is, the resulting change in solid angle is

$$\delta\Omega = -\oint_C \frac{(-d\mathbf{x} \times d\mathbf{l}') \cdot (\mathbf{x} - \mathbf{x}')}{|\mathbf{x} - \mathbf{x}'|^3} \,, \tag{2.110}$$

which can be rewritten in the form

$$\delta\Omega = d\mathbf{x} \cdot \oint_C \frac{d\mathbf{l}' \times (\mathbf{x} - \mathbf{x}')}{|\mathbf{x} - \mathbf{x}'|^3} \,. \tag{2.111}$$

To get from Equation (2.110) to Equation (2.111), we have used the following manipulation involving the permutation tensor ε_{ijk}, which has a value $+1$ if (ijk) is a cyclic permutation of the sequence (123), is equal to -1 if not, and is 0 if 2 or more indices are the same. According to the usual convention, repeated indices are summed:

$$\begin{aligned}
d\mathbf{x} \times d\mathbf{l}' \cdot (\mathbf{x} - \mathbf{x}') &= (\varepsilon_{ijk}\,dx_j\,dl'_k)(x_i - x'_i) \\
&= -\varepsilon_{ijk}\,dl'_k(x_i - x'_i)\,dx_j \\
&= \varepsilon_{jki}\,dl'_k(x_i - x'_i)\,dx_j \\
&= [d\mathbf{l}' \times (\mathbf{x} - \mathbf{x}')] \cdot d\mathbf{x} \,. \tag{2.112}
\end{aligned}$$

Thus, since $\delta\Omega = \vec{\nabla}\Omega \cdot d\mathbf{x}$, we see that

$$\vec{\nabla}\Omega = \oint_C \frac{d\mathbf{l}' \times (\mathbf{x} - \mathbf{x}')}{|\mathbf{x} - \mathbf{x}'|^3} \,, \tag{2.113}$$

and so

$$\mathbf{B} = \frac{I}{c}\vec{\nabla}\Omega \,, \tag{2.114}$$

establishing the fact that Φ_m is proportional to the solid angle subtended at the field point by the source loop. This concept is fully consistent with the requirement of closed current loops in magnetostatics (see § 2.2 above).

Example 2.5. To apply the magnetic scalar potential to a given problem, we must know the solution to Laplace's equation on either side of the prescribed boundary. Let us consider a surface current density

$$\mathbf{K}_S = K(\theta)\hat{\phi} = K_0 \sin\theta\,\hat{\phi} \tag{2.115}$$

distributed on the surface of a sphere. Note that this is not the usual current density $\mathbf{J}$ which has units of current per unit area. Instead, $\mathbf{K}_S$ is the current per unit length. A current such as this might arise in several different situations, like the ones depicted below in Figure 2.9. Three specific examples are (i) a rotating, charged sphere, with surface charge density σ, for which $K_0 = \sigma\omega a$, (ii) a wire-wound sphere, and (iii) a uniformly magnetized sphere, whose net magnetization is equivalent to the surface current density given in Equation (2.115).

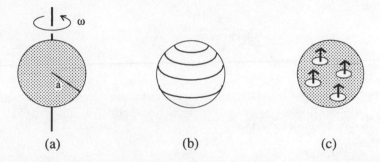

(a) (b) (c)

Figure 2.9 Three equivalent situations for which the technique described in this example will provide a solution. The first (a) is a rotating, uniformly charged sphere; the second (b) is a spherical shell wrapped with appropriately positioned, current-carrying wires; and the third (c) is a uniformly magnetized, stationary sphere.

The general axisymmetric solution of Laplace's equation for Φ_m in spherical coordinates is

$$\Phi_m = \frac{1}{r}\sum_{l=0}^{\infty}\left[a_l r^{l+1} + b_l r^{-l}\right] P_l(\cos\theta)\,. \tag{2.116}$$

In the interior of the sphere, Φ_m is not permitted to be singular at $r = 0$, so that for this region (labeled 1), $b_l = 0$ for all l. In the exterior (labeled 2), Φ_m must converge as $r \to \infty$, so in this region, $a_l = 0$. Thus,

$$\Phi_{m1} = \sum_{l=0}^{\infty} a_l r^l P_l(\cos\theta)\,, \tag{2.117}$$

and

$$\Phi_{m2} = \sum_{l=2}^{\infty} b_l r^{-(l+1)} P_l(\cos\theta) .$$ (2.118)

The surface boundary conditions may be obtained by using the usual Gaussian pillbox with Equation (2.101), and the circuital law with Equation (2.102). Let us first integrate (2.101) over a cylindrical volume straddling the spherical boundary, in such a way that the two faces of the cylinder are parallel to the surface as its height is taken to be vanishingly small. Then, with $\hat{n}$ denoting the outward unit normal vector, we get

$$\int_{\text{Cyl}} \vec{\nabla} \cdot \mathbf{B} = \oint_{\text{Cyl}} \mathbf{B} \cdot \hat{n} \, da = 0 ,$$ (2.119)

which immediately yields

$$(\mathbf{B}_2 - \mathbf{B}_1) \cdot \hat{n} = 0 .$$ (2.120)

Next, let us integrate Equation (2.102) (though with $\mu = 1$) around a rectangular loop C with side lengths h and L in the $\hat{n}$ and $\hat{\theta}$ directions, respectively. We will be taking $h \to 0$ so that only the contribution to the loop integral from the segments parallel to the spherical surface will survive. Then, defining S to be the surface bounded by this rectangle,

$$\oint_S \vec{\nabla} \times \mathbf{B} \cdot (\hat{n} \times \hat{\theta}) \, da = \frac{4\pi}{c} \oint_S \mathbf{J} \cdot (\hat{n} \times \hat{\theta}) \, da .$$ (2.121)

That is,

$$\oint_C \mathbf{B} \cdot d\mathbf{l} = \frac{4\pi}{c} \oint_S \mathbf{J} \cdot (\hat{n} \times \hat{\theta}) \, da ,$$ (2.122)

which becomes

$$L(\mathbf{B}_2 - \mathbf{B}_1) \cdot \hat{\theta} = \frac{4\pi}{c} (\mathbf{K}_S L) \cdot \hat{\phi} ,$$ (2.123)

where $\mathbf{K}_S L$ is the total surface current enclosed by C. Thus, our second boundary condition, complementary to Equation (2.120), is

$$(\mathbf{B}_2 - \mathbf{B}_1) \cdot \hat{\theta} = \frac{4\pi}{c} \mathbf{K}_S \cdot \hat{\phi} .$$ (2.124)

With this, we're almost done, since the expansion coefficients in Equations (2.117) and (2.118) are solved by substituting

$$\mathbf{B} \cdot \hat{n} = B_r = -\frac{\partial \Phi_m}{\partial r} \qquad \text{and} \qquad \mathbf{B} \cdot \hat{\theta} = B_\theta = -\frac{1}{r}\frac{\partial \Phi_m}{\partial \theta}$$ (2.125)

into Equations (2.120) and (2.124). The algebra is straightforward, though somewhat tedious, and results in the solution

$$\Phi_{m1} = -\frac{8\pi K_0}{3c} r \cos\theta , \qquad (2.126)$$

and

$$\Phi_{m2} = \frac{4\pi a^3 K_0}{3c} \frac{\cos\theta}{r^2} . \qquad (2.127)$$

And what does the magnetic induction look like for this configuration of currents? According to the definition of Φ_m,

$$\mathbf{B} = -\vec{\nabla}\Phi_m = -\frac{\partial \Phi_m}{\partial r} \hat{r} - \frac{1}{r}\frac{\partial \Phi_m}{\partial \theta} \hat{\theta} - \frac{1}{r\sin\theta}\frac{\partial \Phi_m}{\partial \phi} \hat{\phi} , \qquad (2.128)$$

and so

$$\mathbf{B}_1 = \frac{8\pi K_0}{3c} \left(\cos\theta\,\hat{r} - \sin\theta\,\hat{\theta} \right) , \qquad (2.129)$$

and

$$\mathbf{B}_2 = \frac{4\pi a^3 K_0}{3c} \left(\frac{2\cos\theta}{r^3} \hat{r} + \frac{\sin\theta}{r^3} \hat{\theta} \right) . \qquad (2.130)$$

Inside the sphere, the field is uniform and has a magnitude $B_z = 8\pi K_0/3c$. Not surprisingly, the field outside is that due to a dipole with a moment $4\pi a^3 K_0/3c$.

2.2.2 Method 2: The Magnetic Vector Potential

Although clearly powerful, the scalar potential method described above is excluded from a large class of problems in which the current density $\mathbf{J}$ is not zero. With a source present, one might immediately think of the Poisson equation, but the situation in magnetostatics is somewhat more involved because $\mathbf{J}$ has three components, unlike the single function source ρ in electrostatics. The idea is to solve for each of these components separately, if we can. According to Equation (1.35), there exists a vector function $\mathbf{A}(\mathbf{x})$ such that $\mathbf{B} = \vec{\nabla} \times \mathbf{A}$, and following the discussion in § 1.3.2, this leads in the Coulomb gauge to the set of three separate equations in (1.74). These are Poisson equations, and so the techniques developed earlier in § 2.1 should yield a complete solution for each of the three independent vector components A_i, and hence the magnetic field through Equation (1.35).

2.2.3 Method 3: Hard Ferromagnets

Interesting new physics is introduced to the problem when the medium has a magnetization $\mathbf{M}(\mathbf{x})$. In general, $\mathbf{M}$ can vary with the applied fields, but this type of situation requires a technical analysis outside the scope of this book. Let us instead consider the case where $\mathbf{M}$ is essentially independent of the external conditions, and in which $\mathbf{J} = 0$. One possible approach is to reintroduce the magnetic scalar potential into Equation (2.101) using (1.50):

$$\vec{\nabla} \cdot \mathbf{B} = \vec{\nabla} \cdot (\mathbf{H} + 4\pi \mathbf{M}) = 0 \, , \tag{2.131}$$

so that

$$\vec{\nabla}^2 \Phi_m = -4\pi \, \rho_m \, . \tag{2.132}$$

Here, we have introduced the effective magnetic charge density (not to be confused with a "real" magnetic charge)

$$\rho_m \equiv -\vec{\nabla} \cdot \mathbf{M} \, . \tag{2.133}$$

The result is the Poisson equation with a fictitious source ρ_m, but which is nonetheless amenable to the methods of solution discussed earlier.

On the surface of a substance with $\mathbf{M} \neq 0$, ρ_m results in an effective magnetic surface charge density σ_m. To see how this comes about, consider again the application of the divergence theorem to a Gaussian pillbox with cross-sectional area a straddling the boundary S between the magnetization region and vacuum. Following the same procedure as we did in § 2.2.1, we set up the integration over the volume V of the pillbox and through the divergence theorem determine the surface condition:

$$\begin{aligned}
\int_V \rho_m \, d^3x &= -\int_V \vec{\nabla} \cdot \mathbf{M} \, d^3x \\[2mm]
&= -\oint_S \mathbf{M} \cdot \hat{n} \, da \\[2mm]
&= -\mathbf{M} \cdot \hat{n} \, a \, .
\end{aligned} \tag{2.134}$$

As the volume of the pillbox is made vanishingly small, $\int \rho_m \, d^3x$ approaches the limit $\sigma_m \, a$, resulting in the identity

$$\sigma_m = \hat{n} \cdot \mathbf{M} \, . \tag{2.135}$$

Problems such as this can also be handled using the vector potential, since

$$\vec{\nabla} \times \mathbf{H} = \vec{\nabla} \times (\mathbf{B} - 4\pi \mathbf{M}) = 0 \, , \tag{2.136}$$

Differential	Integral
$\vec{\nabla} \cdot \mathbf{B} = 0$	$\oint_S \mathbf{B} \cdot \hat{n}\,da = 0$
$\vec{\nabla} \times \mathbf{B} = 4\pi \mathbf{J}/c$	$\oint_C \mathbf{B} \cdot d\mathbf{l} = 4\pi I/c$
$\vec{\nabla}^2 \Phi_m = 0$	$\Phi_m = -I\Omega/c$
$\vec{\nabla}^2 \mathbf{A} = -4\pi \mathbf{J}/c$	$\mathbf{A} = \int d^3x'\, \mathbf{J}(\mathbf{x}')/c|\mathbf{x} - \mathbf{x}'|$

Table 2.2 Key relations in magnetostatics: the differential and integral properties of $\mathbf{B}$, Φ_m, and the vector potential $\mathbf{A}$.

which leads immediately to

$$\vec{\nabla}^2 \mathbf{A} = -\frac{4\pi}{c} \mathbf{J}_m \,. \tag{2.137}$$

Instead of a fictitious magnetic charge density ρ_m, this approach introduces an effective magnetic current density

$$\mathbf{J}_m \equiv c\vec{\nabla} \times \mathbf{M} \,, \tag{2.138}$$

but the idea is the same. This definition reduces the pertinent equations to a set of three Poisson equations for the components of $\mathbf{A}$, and the method of solution then follows the various steps outlined above.

We close this section, and this chapter, by summarizing (in Table 2.2) the key relations we have encountered in magnetostatics. These constitute the differential and integral properties of $\mathbf{B}$, Φ_m, and the vector potential $\mathbf{A}$. Note that the integral form of the first equation in this table is valid only for a closed surface. Only then do we have the requirement that every field line entering this surface must also exit somewhere else (see the discussion preceding Equation [1.14]).

3

GENERAL PROPERTIES OF MAXWELL'S EQUATIONS

3.1 TIME-VARYING FIELDS

The independence of the electric and magnetic fields disappears when the charge density ρ and current density $\mathbf{J}$ are no longer in steady state. Indeed, temporal variations lead to the surprising property that fields can be sources for each other. In this chapter, we will be taking the first steps toward unifying electric and magnetic phenomena, a process that will culminate with our development of a relativistic, field theoretic formulation. Rest assured, however, that the equations we derive here for the fields (indeed, starting with Maxwell's equations) are correct and complete, since they incorporate all the necessary physics pertaining to the frame in which the fields are calculated and measured. What we shall find in Chapter 5 is that the introduction of special relativity affects the dynamics of particles interacting with the fields, and the manner in which $\mathbf{E}$ and $\mathbf{B}$ transform from frame to frame. Very importantly, casting the field equations into the language of special relativity will give us a better insight into their underlying (unified) nature.

Since Faraday's law (Equation [1.16]) deals empirically with the electric field induced by a changing magnetic flux, it is a natural place for us to begin. Right away we need to be careful about distinguishing between physical quantities measured in different reference frames, because the time derivative includes both an intrinsic time piece and a convective term due to the relative motion of the observer:

$$\frac{d}{dt} = \frac{\partial}{\partial t} + \mathbf{v} \cdot \vec{\nabla} \ . \tag{3.1}$$

In the laboratory frame, the right-hand side of (1.16) is simply

$$-\frac{1}{c}\frac{d\Phi_B}{dt} = -\frac{1}{c}\left(\frac{\partial}{\partial t} + \mathbf{v}\cdot\vec{\nabla}\right)\int_S \mathbf{B}\cdot\hat{n}\,da = -\frac{1}{c}\int_S \frac{\partial\mathbf{B}}{\partial t}\cdot\hat{n}\,da\,, \qquad (3.2)$$

since $\mathbf{B}$ is the field measured by an observer at rest ($\mathbf{v} = 0$) in this frame. However, in the primed (i.e., rest) frame of reference attached to the *rigid* loop C, which is itself moving with velocity $\mathbf{v}$ relative to the laboratory, Faraday's law should correctly be written

$$\oint_C \mathbf{E}'\cdot d\mathbf{l}' = -\frac{1}{c}\int_S \left(\frac{\partial}{\partial t} + \mathbf{v}\cdot\vec{\nabla}\right)\mathbf{B}\cdot\hat{n}\,da\,, \qquad (3.3)$$

since the derivative is now taken in a frame moving with velocity $\mathbf{v}$ relative to the observer who determines that the magnetic field intensity is $\mathbf{B}$. Note that although we are taking this relative motion into account, this is nonetheless an entirely nonrelativistic effect. We have not yet introduced special relativity (see Chapter 5), so no "transformation" of the time or $\mathbf{B}$ is to be made at this point of our discussion. Now,

$$(\mathbf{v}\cdot\vec{\nabla})\mathbf{B} = \vec{\nabla}\times(\mathbf{B}\times\mathbf{v}) + \mathbf{v}(\vec{\nabla}\cdot\mathbf{B}) - \mathbf{B}(\vec{\nabla}\cdot\mathbf{v}) + (\mathbf{B}\cdot\vec{\nabla})\mathbf{v}\,, \qquad (3.4)$$

so that with $\vec{\nabla}\cdot\mathbf{B} = 0$ (and spatial derivatives of $\mathbf{v}$ being zero),

$$(\mathbf{v}\cdot\vec{\nabla})\mathbf{B} = \vec{\nabla}\times(\mathbf{B}\times\mathbf{v})\,. \qquad (3.5)$$

Thus,

$$\oint_C \mathbf{E}'\cdot d\mathbf{l}' = -\frac{1}{c}\int_S \frac{\partial\mathbf{B}}{\partial t}\cdot\hat{n}\,da + \frac{1}{c}\oint_C \mathbf{v}\times\mathbf{B}\cdot d\mathbf{l}\,. \qquad (3.6)$$

The combination of Equations (3.2) and (3.6) leads to the identification

$$\boxed{\mathbf{E}' = \mathbf{E} + (\mathbf{v}\times\mathbf{B})/c}\,, \qquad (3.7)$$

that is, we interpret the velocity-dependent term as the transformed change in $\mathbf{E}$.

Maxwell's equations have built into them the empirical fact that time derivatives of the fields are sources for each other's counterpart. The result exhibited here is perhaps a little stronger, in the sense that one field is seen to "transform" into the other by virtue of a relative motion. That which we see as the effect of a magnetic field in the laboratory would be interpreted as the result of an interaction between a charge and an "electric" field $\mathbf{v}\times\mathbf{B}/c$ by an observer in the moving frame. This is the first hint of an underlying union of electric and magnetic phenomena that we will address more fully in Chapter 5 and beyond.

3.2   THE TIME-DEPENDENT GREEN FUNCTION

We can learn more about the time-dependent nature of our theory by also considering the behavior of the scalar and vector potentials. The dynamic evolution of Φ and $\mathbf{A}$ is determined by the inhomogeneous Maxwell equations that contain the sources. With the dependence of $\mathbf{E}$ on Φ and $\mathbf{A}$ as shown in Equation (1.37), Gauss's law (1.12) transforms to

$$\vec{\nabla}^2 \Phi + \frac{1}{c}\frac{\partial}{\partial t}\left(\vec{\nabla}\cdot\mathbf{A}\right) = -4\pi\rho\,. \tag{3.8}$$

This is a scalar equation containing four unknowns, Φ, and the three components of $\mathbf{A}$. The other essential equations follow from Ampère's law (1.34), which yields the following expression for the potentials once we substitute for $\mathbf{E}$ and $\mathbf{B}$:

$$\vec{\nabla}^2 \mathbf{A} - \frac{1}{c^2}\frac{\partial^2 \mathbf{A}}{\partial t^2} - \vec{\nabla}\left(\vec{\nabla}\cdot\mathbf{A} + \frac{1}{c}\frac{\partial\Phi}{\partial t}\right) = -\frac{4\pi}{c}\mathbf{J}\,. \tag{3.9}$$

Equations (3.8) and (3.9) constitute a complete set of four equations in four unknowns, which in principle can be solved to get the full spatial and temporal dependence of the potentials, and eventually the fields through (1.35) and (1.37). As they stand, these expressions couple all the potentials strongly, a consequence of the interdependence of $\mathbf{E}$ and $\mathbf{B}$ when we allow for nonsteady situations. However, there is a way to restructure these relations with the appropriate choice of a gauge. We have already introduced this concept in § 1.3.2, where it was shown that the natural gauge to choose in a time-independent setting is the Coulomb gauge defined by the condition $\vec{\nabla}\cdot\mathbf{A} = 0$. This clearly would not work here since $\partial\mathbf{A}/\partial t \neq 0$. Adding a term like $\vec{\nabla}\psi$ to $\mathbf{A}$ (see Equation [1.71]) would still leave $\mathbf{B}$ unaffected, but not $\mathbf{E}$, since it is unclear whether or not $\partial/\partial t(\vec{\nabla}\psi)$ should always be zero.

Any gauge transformation must therefore include both space and time. For example, we can decouple Equations (3.8) and (3.9) by making the potential transformations

$$\mathbf{A} \to \mathbf{A} + \vec{\nabla}\psi\,, \tag{3.10}$$

and

$$\Phi \to \Phi - \frac{1}{c}\frac{\partial\psi}{\partial t}\,, \tag{3.11}$$

as long as

$$\vec{\nabla}\cdot\mathbf{A} + \frac{1}{c}\frac{\partial\Phi}{\partial t} = 0\,, \tag{3.12}$$

which allows us to replace (3.8) and (3.9) with the set

$$\vec{\nabla}^2\Phi - \frac{1}{c^2}\frac{\partial^2\Phi}{\partial t^2} = -4\pi\rho \, , \tag{3.13}$$

and

$$\vec{\nabla}^2\mathbf{A} - \frac{1}{c^2}\frac{\partial^2\mathbf{A}}{\partial t^2} = -\frac{4\pi}{c}\mathbf{J} \, . \tag{3.14}$$

It is straightforward to see that a function ψ can always be found to satisfy this condition. Suppose that in fact a given choice of potentials $\mathbf{A}$ and Φ do not satisfy Equation (3.12). A gauge-transformed set $\mathbf{A}'$ and Φ' then clearly will satisfy this condition as long as $\mathbf{A}' = \mathbf{A} + \vec{\nabla}\psi$ and $\Phi' = \Phi - \partial\psi/\partial ct$, with $\vec{\nabla}^2\psi - \partial^2\psi/\partial(ct)^2 = -(\vec{\nabla}\cdot\mathbf{A} + \partial\Phi/\partial ct)$. In addition, this also shows that if $\mathbf{A}$ and Φ initially satisfy Equation (3.12), any gauge transformation with $\vec{\nabla}^2\psi - \partial^2\psi/\partial(ct)^2 = 0$ will preserve this condition.

Does this mean we have found a way to make Φ and $\mathbf{A}$ (and hence possibly $\mathbf{E}$ and $\mathbf{B}$) independent of each other again? Of course not. Although Equations (3.13) and (3.14) can be solved separately for the four potential components, they are nonetheless always linked through the gauge relation (3.12). The choice of ψ that results in Equation (3.12) is known as the *Lorenz* gauge transformation, and Equations (3.13) and (3.14) are the wave equations for Φ and $\mathbf{A}$ in the Lorenz gauge. (Note that this was the Danish physicist Ludvig V. Lorenz, not the Dutch physicist Hendrik A. Lorentz whose name is attached to many other physical laws and ideas described in this book; see Van Bladel 1991.) We shall learn a great deal more about the physical meaning of this gauge invariance of the electromagnetic field in § 6.4, but we can immediately get a feeling for its origin by noting that when we add the partial time derivative of Equation (3.13) to the divergence of Equation (3.14), Equation (3.12) reduces the result to the charge continuity equation (1.32). That is, a gauge condition, such as the Lorenz relation (3.12), applied to the potential wave equations yields the conservation of charge.

It is worthwhile pausing for a moment to think about yet another indication emerging from the gauge invariance of the electromagnetic field—that there must exist an underlying union of electric and magnetic phenomena. It is not a coincidence that the "natural" gauge for the potentials is one in which *four* potential components are coupled using coordinates in *four* dimensions. We had no a priori indication that it had to be this way, but as we shall see later, this assemblage points to the existence of a more elaborate infrastructure known as four-dimensional spacetime.

We solve these equations using a generalized time-dependent Green function. One way of facilitating this procedure is to first separate out the spatial and time dependences by writing Φ and $\mathbf{A}$ as superpositions of functions at different frequencies, i.e., by expanding Φ and $\mathbf{A}$ as Fourier series or integrals. As long as a function $g(t)$ satisfies the conditions (1) that both $g(t)$ and $dg(t)/dt$ are piecewise continuous in every finite interval of t, and (2) that $\int_{-\infty}^{\infty} |g(t)|\, dt$ converges, i.e., that $g(t)$ is absolutely integrable in $(-\infty, \infty)$, the Fourier theorem assures us that any such function can be expanded in this fashion. It also happens that often a handful of frequencies (sometimes just one) dominate the behavior of the solutions. We thus write

$$ f(\mathbf{x}, t) = \frac{1}{2\pi} \int_{-\infty}^{\infty} \tilde{f}(\mathbf{x}, \omega) \, \exp(-i\omega t) \, d\omega \, , \tag{3.15} $$

where $f(\mathbf{x}, t)$ is one of either Φ, ρ, or the components of $\mathbf{A}$ and $\mathbf{J}$, and $\tilde{f}(\mathbf{x}, \omega)$ is its Fourier transform. Equations (3.13) and (3.14) thereby reduce to

$$ \left[\vec{\nabla}^2 + \left(\frac{\omega}{c} \right)^2 \right] \tilde{\Phi}(\mathbf{x}, \omega) = -4\pi \, \tilde{\rho}(\mathbf{x}, \omega) \, , \tag{3.16} $$

and

$$ \left[\vec{\nabla}^2 + \left(\frac{\omega}{c} \right)^2 \right] \tilde{\mathbf{A}}(\mathbf{x}, \omega) = \frac{4\pi}{c} \, \tilde{\mathbf{J}}(\mathbf{x}, \omega) \, , \tag{3.17} $$

since the integral expressions must be valid for all t. These are elliptic partial differential equations similar to the Poisson equation. Thus, one way of solving them is to use a Green function together with Green's identities, which we derived from the divergence theorem. Following the ideas developed in § 2.1.2 above, we seek a function $\tilde{G}_\omega(\mathbf{x} - \mathbf{x}')$ such that

$$ \left[\vec{\nabla}^2 + \left(\frac{\omega}{c} \right)^2 \right] \tilde{G}_\omega(\mathbf{x} - \mathbf{x}') = -4\pi \, \delta^3(\mathbf{x} - \mathbf{x}') \, . \tag{3.18} $$

Since, moreover, $\tilde{G}_\omega$ is a function only of $|\mathbf{x} - \mathbf{x}'|$, not the absolute orientation of the vector $\mathbf{x} - \mathbf{x}'$,

$$ \frac{1}{r} \frac{d^2}{dr^2} \left(r\tilde{G}_\omega \right) + \left(\frac{\omega}{c} \right)^2 \tilde{G}_\omega = -4\pi \, \delta(r) \, , \tag{3.19} $$

where $r \equiv |\mathbf{x} - \mathbf{x}'|$.

The solution to this equation is

$$ \tilde{G}_\omega^\pm(r) = \frac{\exp(\pm ikr)}{r} \, , \tag{3.20} $$

as may be readily verified by substitution into (3.19). Here, $k \equiv \omega/c$, and we note that the δ-function arises from the normalization at $r = 0$, as it did in § 2.1.2 for the electrostatic field. Thus, the time-dependent Green function is

$$G^{\pm}(r, t) = \frac{1}{2\pi} \int_{-\infty}^{\infty} \frac{\exp(\pm ikr)}{r} \exp(-i\omega t)\, d\omega \,, \qquad (3.21)$$

where t is measured relative to a prescribed source time. That is, if the source activity is known at some time t', then the resulting Green function at time t is

$$G^{\pm}(\mathbf{x}, t; \mathbf{x}', t') = \frac{1}{2\pi} \int_{-\infty}^{\infty} \frac{\exp(\pm ik|\mathbf{x} - \mathbf{x}'|)}{|\mathbf{x} - \mathbf{x}'|} \exp(-i\omega[t - t'])\, d\omega \,. \qquad (3.22)$$

We recognize the right-hand side of this equation as the Dirac delta function, and so we arrive at the compact form for G written as

$$\boxed{G^{\pm}(r, t - t') = \delta(t - t' \mp r/c)/r} \,. \qquad (3.23)$$

This is a wonderful result. G^{+} is the *retarded* Green function because it exhibits the *causal* behavior associated with a wave disturbance traveling with speed c. An effect observed at $(\mathbf{x}, t)$ was caused by the action of a source at $(\mathbf{x}', t - r/c)$. Similarly, G^{-} is the *advanced* Green function, relating a potential field observed at $(\mathbf{x}, t)$ to the activity of a source at $(\mathbf{x}', t + r/c)$. How this comes about will be the subject of the following discussion.

In § 2.1.2, we learned that when the surface of the domain of solution is pushed to infinity, only the internal source term contributes to the potential. The main difference between that time-independent derivation and the situation we have here is that one now needs to include all contributions to the potential from sources acting at possibly many different times. The only restriction is that the difference between the source time and the observation (or field) time should equal the light travel time between the two points. Thus, a source that is very far away from the observation point could have acted much longer in the past and still contribute to the measured potential, than a source that is closer, which acted more recently. Mathematically, this means that our potentials are now double integrals, over both source space and time:

$$\Phi^{\pm}(\mathbf{x}, t) = \int \int G^{\pm}(\mathbf{x}, t; \mathbf{x}', t')\, \rho(\mathbf{x}', t')\, d^3 x'\, dt' \,, \qquad (3.24)$$

and

$$\mathbf{A}^{\pm}(\mathbf{x}, t) = \frac{1}{c} \int \int G^{\pm}(\mathbf{x}, t; \mathbf{x}', t')\, \mathbf{J}(\mathbf{x}', t')\, d^3 x'\, dt' \,. \qquad (3.25)$$

When we know the incoming waves $\Phi_{in}(\mathbf{x}, t)$ and $\mathbf{A}_{in}(\mathbf{x}, t)$ and we wish to determine the overall potentials taking into account the augmentation due to the action of ρ and $\mathbf{J}$, then the appropriate solutions to use are the retarded ones. In other words, the action of the sources adds to the initial conditions to produce the current potentials, so

$$\Phi(\mathbf{x}, t) = \Phi_{in}(\mathbf{x}, t) + \int \int_{t' < t} G^+(\mathbf{x}, t; \mathbf{x}', t')\, \rho(\mathbf{x}', t')\, d^3x'\, dt' , \qquad (3.26)$$

$$\mathbf{A}(\mathbf{x}, t) = \mathbf{A}_{in}(\mathbf{x}, t) + \frac{1}{c} \int \int_{t' < t} G^+(\mathbf{x}, t; \mathbf{x}', t')\, \mathbf{J}(\mathbf{x}', t')\, d^3x'\, dt' . \qquad (3.27)$$

It must be stressed here that Φ_{in} and $\mathbf{A}_{in}$ satisfy the homogeneous (i.e., source-less) equations, which accounts for the fact that as $\mathbf{x}$ and t change, so too do these "initial" potentials evolve. That is, the argument of these functions should correctly be $\mathbf{x}$ and t, rather than $\mathbf{x}_{in}$ and t_{in}.

When instead the outgoing solutions $\Phi_{out}(\mathbf{x}, t)$ and $\mathbf{A}_{out}(\mathbf{x}, t)$ are given, the potentials measured now are the backward-evolved outgoing functions, minus any augmentation due to the sources between now and $t' \to \infty$. That is, the appropriate solutions to use in this circumstance are the advanced ones:

$$\Phi(\mathbf{x}, t) = \Phi_{out}(\mathbf{x}, t) + \int \int_{t' > t} G^-(\mathbf{x}, t; \mathbf{x}', t')\, \rho(\mathbf{x}', t')\, d^3x'\, dt' , \qquad (3.28)$$

and

$$\mathbf{A}(\mathbf{x}, t) = \mathbf{A}_{out}(\mathbf{x}, t) + \frac{1}{c} \int \int_{t' > t} G^-(\mathbf{x}, t; \mathbf{x}', t')\, \mathbf{J}(\mathbf{x}', t')\, d^3x'\, dt' . \qquad (3.29)$$

As long as we know the source functions $\rho(\mathbf{x}', t')$ and $\mathbf{J}(\mathbf{x}', t')$, these expressions account for the potentials correctly even in the relativistic domain. However, when the particles are moving rapidly, relativistic effects must be included in order to adequately handle the radiative motion of the sources. This is why we shall revisit this topic in Chapter 7, when the necessary relativistic tools will have been developed. We shall find there that $\rho(\mathbf{x}', t')$ and $\mathbf{J}(\mathbf{x}', t')$ themselves can depend on $\Phi(\mathbf{x}, t)$ and $\mathbf{A}(\mathbf{x}, t)$, since the energy and momentum carried away by the field induce a change in the particle's trajectory. These effects become important when the velocity of the charge carriers approaches c, so relativistic mechanics will broaden the useful range of these expressions; even without these considerations, however, there exists a large class of problems for which the present formulation is entirely valid. In going through this exercise of analyzing the nonrelativistic aspects of time-dependent fields, we are learning that electromagnetic theory as it stands is somewhat incomplete, not because the fields and their potentials are not known precisely, but rather because the description of the sources that give rise to them is to this point still couched in the language of nonrelativistic mechanics.

3.3   CONSERVATION LAWS

It was pointed out in the Introduction that the field concept became firmly entrenched after the development of Maxwell's equations, when it was realized that the field is more than a mathematical device. Carrying energy, and linear and angular momentum, the electric and magnetic fields are dynamical entities, on par with the sources that produce and interact with them. We have said nothing yet about the fact that the constituents of these fields are now known to be photons, which of course are particles, and for the ideas developed in this section we don't need to. We will see that the fields are imbued with these dynamical characteristics irrespective of their internal structure.

3.3.1   Field Energy Density and Poynting's Theorem

To begin with, we know at the most basic level using Coulomb's empirical equation that to assemble a cluster of charges we must do work on the system. It is not surprising, therefore, that a charge distribution is associated with an energy. But this energy is described in terms of the charges and is therefore interpreted as being coupled directly to them. To show that the energy is in fact stored in the fields, we should be able to account for it by using the properties of the fields themselves.

Suppose we bring an element of charge density $\delta\rho(\mathbf{x})$ in from infinity to the coordinate point $\mathbf{x}$, where the potential is known to be $\Phi(\mathbf{x})$. (In this section, we will use a prime to denote the instantaneous value of quantities that are changing as the system is built up with charge.) The energy density of the system will then change by an amount

$$\delta w(\mathbf{x}) = \delta\rho(\mathbf{x})\,\Phi(\mathbf{x})\;. \tag{3.30}$$

Thus, over the whole volume V of interest,

$$\delta W = \int_V \delta w(\mathbf{x})\,d^3x = \int_V \delta\rho(\mathbf{x})\,\Phi(\mathbf{x})\,d^3x\;. \tag{3.31}$$

This is the description of the system's binding energy in terms of the charges. But we know from Gauss's law that

$$\delta\rho(\mathbf{x}) = \frac{1}{4\pi}\vec{\nabla}\cdot(\delta\mathbf{E}) , \qquad (3.32)$$

where $\delta\mathbf{E}$ is the corresponding change in $\mathbf{E}$ due to the local change in source density $\delta\rho(\mathbf{x})$. Thus, using the divergence theorem with the enclosing surface S, we see that

$$
\begin{aligned}
\delta W &= \int_V \frac{1}{4\pi}\vec{\nabla}\cdot[\delta\mathbf{E}(\mathbf{x})]\,\Phi(\mathbf{x})\,d^3x \\
&= \int_V \frac{1}{4\pi}\left\{\vec{\nabla}\cdot[\delta\mathbf{E}(\mathbf{x})\,\Phi(\mathbf{x})] - \delta\mathbf{E}(\mathbf{x})\cdot\vec{\nabla}\Phi(\mathbf{x})\right\}\,d^3x \\
&= \frac{1}{4\pi}\int_S (\delta\mathbf{E}\,\Phi)\cdot\hat{n}\,da - \frac{1}{4\pi}\int_V \delta\mathbf{E}\cdot\vec{\nabla}\Phi\,d^3x \\
&= \frac{1}{4\pi}\int_V \delta\mathbf{E}\cdot\mathbf{E}\,d^3x \\
&= \frac{1}{4\pi}\delta\int_V \frac{1}{2}(\mathbf{E}\cdot\mathbf{E})\,d^3x .
\end{aligned}
\qquad (3.33)
$$

The last step follows from the definition of Φ in terms of $\mathbf{E}$ and the fact that S can be taken arbitrarily large whereas $\delta\rho$ must remain localized. The total electrostatic energy of the system is therefore

$$W = \frac{1}{8\pi}\int_V \mathbf{E}\cdot\mathbf{E}\,d^3x . \qquad (3.34)$$

Notice that this expression for the system's binding energy makes no mention at all of the assembled charges but is instead written entirely in terms of the electric field intensity. From it, we infer that the energy density of an electrostatic configuration must be

$$\boxed{u_E = |\mathbf{E}|^2/8\pi} . \qquad (3.35)$$

For a magnetostatic configuration, the corresponding change in energy of the system due to a change in the position $\delta\mathbf{r}$ of a loop element $d\mathbf{l}$ carrying a current $I(\mathbf{x})$ is

$$\delta W(\mathbf{x}) = -\frac{I}{c}\,[d\mathbf{l}\times\mathbf{B}(\mathbf{x})]\cdot\delta\mathbf{r} , \qquad (3.36)$$

where $\mathbf{B}(\mathbf{x})$ is the instantaneous magnetic field intensity at the location of the loop. This is simply the work done against the force $I \, d\mathbf{l} \times \mathbf{B}(\mathbf{x})/c$. Thus, using the same vector manipulation that led to Equation (2.111) earlier, we have

$$\delta W(\mathbf{x}) = \frac{I}{c} \, (d\mathbf{l} \times \delta\mathbf{r}) \cdot \mathbf{B}(\mathbf{x}) , \qquad (3.37)$$

and noting that $d\mathbf{l} \times \delta\mathbf{r} = -\hat{n} \, \delta a$ (the inward-pointing elemental area),

$$\delta W(\mathbf{x}) = -\frac{I}{c} \, \delta\Phi_B(\mathbf{x}) , \qquad (3.38)$$

where $\delta\Phi_B(\mathbf{x}) = \hat{n} \, \delta a \cdot \mathbf{B}(\mathbf{x})$ is an element of magnetic flux threading the loop. Thus, replacing I with the field it produces results in the following identification of the magnetic field energy density:

$$\boxed{u_B = |\mathbf{B}|^2/8\pi} \ . \qquad (3.39)$$

Let us now examine what happens to this energy density when $\mathbf{E}$ and $\mathbf{B}$ are time dependent. Suppose both an electric and a magnetic field are present in a volume of space V. Using the above relations, the total electromagnetic energy in the system changes in time according to the expression

$$\frac{\partial}{\partial t} \int_V (u_E + u_B) \, d^3x = \frac{1}{8\pi} \int_V \left(2\mathbf{E} \cdot \frac{\partial \mathbf{E}}{\partial t} + 2\mathbf{B} \cdot \frac{\partial \mathbf{B}}{\partial t} \right) d^3x . \qquad (3.40)$$

That is, replacing $\partial \mathbf{E}/\partial t$ and $\partial \mathbf{B}/\partial t$ with their counterparts from Faraday's and Ampère's laws, respectively,

$$\frac{\partial}{\partial t} \int_V (u_E + u_B) \, d^3x = \frac{1}{4\pi} \int_V \left\{ \mathbf{E} \cdot \left[c\vec{\nabla} \times \mathbf{B} - 4\pi\mathbf{J} \right] + \mathbf{B} \cdot \left[-c\vec{\nabla} \times \mathbf{E} \right] \right\} d^3x . \qquad (3.41)$$

But $-\mathbf{B} \cdot (\vec{\nabla} \times \mathbf{E}) + \mathbf{E} \cdot (\vec{\nabla} \times \mathbf{B}) = -\vec{\nabla} \cdot (\mathbf{E} \times \mathbf{B})$, so that

$$\frac{\partial}{\partial t} \int_V (u_E + u_B) \, d^3x = -\int_V \mathbf{J} \cdot \mathbf{E} \, d^3x + \frac{c}{4\pi} \int_V -\vec{\nabla} \cdot (\mathbf{E} \times \mathbf{B}) \, d^3x , \qquad (3.42)$$

and since the volume is arbitrary, we must therefore have

$$\boxed{\partial u/\partial t + c\vec{\nabla} \cdot (\mathbf{E} \times \mathbf{B})/4\pi = -\mathbf{J} \cdot \mathbf{E}} , \qquad (3.43)$$

where $u \equiv u_E + u_B$. This is the *conservation of energy* equation for the electromagnetic field. The quantity

$$\boxed{\mathbf{S} \equiv c\,(\mathbf{E} \times \mathbf{B})/4\pi} \tag{3.44}$$

has units of energy per unit area per unit time and is known as the Poynting vector. It represents the energy flux out of the volume associated with the enclosed electromagnetic field. The other term, $-\mathbf{J} \cdot \mathbf{E}$, is the work done (mechanically) on the system of charges and is therefore a *dissipation* of electromagnetic energy. Thus, we see that not only does u satisfy an energy conservation equation, but in addition, the electromagnetic energy is seen to be exchangeable with that of the particles with which the field is interacting.

3.3.2   Conservation of Linear Momentum

In a similar fashion, we can consider the linear momentum content of the electromagnetic field. We do this by using the transfer of momentum during an interaction between the field and charges as a probe of its dependence on the characteristics of the field. We know from the application of Newton's second law to the total mechanical momentum $\mathbf{P}_{\text{mech}}$ that

$$\frac{d\mathbf{P}_{\text{mech}}}{dt} = \mathbf{F}_{\text{tot}} , \tag{3.45}$$

where $\mathbf{F}_{\text{tot}}$ is the total force on the particles due to the field. Thus, generalizing Equation (1.3) for a distributed charge,

$$\frac{d\mathbf{P}_{\text{mech}}}{dt} = \int_V \rho \left(\mathbf{E} + \frac{1}{c}\, \mathbf{v} \times \mathbf{B} \right) d^3 x . \tag{3.46}$$

The idea here is to replace all references to the sources ρ and $\mathbf{J} \equiv \rho \mathbf{v}$ by their equivalent forms in terms of the fields. We therefore put $\rho = \vec{\nabla} \cdot \mathbf{E}/4\pi$ and substitute for $\mathbf{J}$ from Equation (1.34). The result is

$$\frac{d\mathbf{P}_{\text{mech}}}{dt} = \frac{1}{4\pi} \int_V \left[\mathbf{E}(\vec{\nabla} \cdot \mathbf{E}) + \frac{1}{c}\mathbf{B} \times \frac{\partial \mathbf{E}}{\partial t} - \mathbf{B} \times (\vec{\nabla} \times \mathbf{B}) \right] d^3 x . \tag{3.47}$$

It is difficult to make headway from here unless we first clearly identify our goals. Ideally, we should be able to extract a term from the right-hand side of (3.47) that looks like the $d\mathbf{P}_{\text{mech}}/dt$ term on the other side. After all, since this equation is set up to give us the rate of change of particle momentum, we can reasonably expect that this momentum should be extracted conservatively from the field. We therefore need to identify from the right-hand side the total time derivative of a quantity that has units of momentum. The middle term looks like it belongs in this category, but it involves the time derivative of only one field, so let's work with it. We will write

$$\mathbf{B} \times \frac{\partial \mathbf{E}}{\partial t} = -\frac{\partial}{\partial t}(\mathbf{E} \times \mathbf{B}) + \mathbf{E} \times \frac{\partial \mathbf{B}}{\partial t} \, , \qquad (3.48)$$

so that now

$$\frac{d\mathbf{P}_{\text{mech}}}{dt} = \frac{1}{4\pi} \int_V \left[\mathbf{E}(\vec{\nabla} \cdot \mathbf{E}) + \frac{1}{c}\mathbf{E} \times \frac{\partial \mathbf{B}}{\partial t} \right.$$

$$\left. - \mathbf{B} \times (\vec{\nabla} \times \mathbf{B}) - \frac{1}{c}\frac{\partial}{\partial t}(\mathbf{E} \times \mathbf{B}) \right] d^3x \, . \qquad (3.49)$$

At first it doesn't look like we've gained much, except that the $\mathbf{E} \times \partial \mathbf{B}/\partial t$ term is really complementary to $\mathbf{B} \times (\vec{\nabla} \times \mathbf{B})$ when we replace $\partial \mathbf{B}/\partial t$ using Faraday's law:

$$\frac{d\mathbf{P}_{\text{mech}}}{dt} = \frac{1}{4\pi} \int_V \left[\mathbf{E}(\vec{\nabla} \cdot \mathbf{E}) - \mathbf{E} \times (\vec{\nabla} \times \mathbf{E}) \right.$$

$$\left. - \mathbf{B} \times (\vec{\nabla} \times \mathbf{B}) - \frac{1}{c}\frac{\partial}{\partial t}(\mathbf{E} \times \mathbf{B}) \right] d^3x \, . \qquad (3.50)$$

More important, we have extracted a term that looks like it has the characteristics we need and the first three terms on the right-hand side are unrelated to the momentum since they do not involve any time derivatives like the other two terms in this equation. For purely aesthetic reasons, we might also want to add a term $\mathbf{B}(\vec{\nabla} \cdot \mathbf{B}) = 0$ to the right-hand side to make the expression completely symmetric in $\mathbf{E}$ and $\mathbf{B}$. Thus, since the integral over d^3x reduces to a function only of t, we can put $\partial/\partial t \to d/dt$ and we get finally

$$\frac{d\mathbf{P}_{\text{mech}}}{dt} + \frac{d}{dt}\int_V \mathbf{g}\, d^3x = \frac{1}{4\pi} \int_V \left[\mathbf{E}(\vec{\nabla} \cdot \mathbf{E}) - \mathbf{E} \times (\vec{\nabla} \times \mathbf{E}) \right.$$

$$\left. + \mathbf{B}(\vec{\nabla} \cdot \mathbf{B}) - \mathbf{B} \times (\vec{\nabla} \times \mathbf{B}) \right] d^3x \, , \qquad (3.51)$$

where

$$\boxed{\mathbf{g} \equiv (\mathbf{E} \times \mathbf{B})/4\pi c} \qquad (3.52)$$

can now be identified as the *electromagnetic momentum density*. The right-hand side of this equation is itself very meaningful and will be the subject of the following subsection.

We have thus derived the remarkable result that

$$\mathbf{g} = \frac{1}{c^2}\,\mathbf{S}\,. \tag{3.53}$$

Knowing that electromagnetic phenomena in vacuum travel with a speed c, it is easy to see that the energy density carried by traveling fields should be $|\mathbf{S}|/c$ since $\mathbf{S}$ is the energy flux. Equation (3.53) is now making an unprecedented advance beyond this by implying that the momentum density in the fields is simply this energy density divided by c. We stress that the derivation of this result has been completely nonrelativistic, but it is nonetheless consistent with the physical insights we shall gain during our study of special relativity. There we will learn that a photon's energy E_ν and its momentum p_ν are related by the expression $E_\nu = c\,p_\nu$, and so Equation (3.53) is merely a restatement (or more accurately, a prestatement) of this result, though here applied to an ensemble of photons constituting the electric and magnetic fields.

3.3.3   The Maxwell Stress Tensor

The right-hand side of Equation (3.51) merits further attention. Based on the nature of the equation and the form of the left-hand side, it clearly represents the integrated momentum flux. In classical particle dynamics, this portion of the equation would therefore be expressible as the gradient of a scalar potential function, i.e., a gradient representation for the force. Given the complicated dependence of this integral on $\mathbf{E}$ and $\mathbf{B}$, it would clearly be difficult to do this here, since for example, terms like $\mathbf{E}(\vec{\nabla}\mathbf{E})$ mix the field components. If we wish to retain this potential gradient aspect of the problem, we need to move to a higher order function—a tensor—and we must then try to guess its form. Equation (3.51) mixes field components, but never more than two at a time. We thus expect that the tensor should be of rank 2, and we will denote it by $T_{\alpha\beta}$, where α and β run over all possible indices $1, 2, 3$. We can get some clues as to its appearance by looking at terms like the following:

$$[\mathbf{E}(\vec{\nabla}\cdot\mathbf{E}) - \mathbf{E}\times(\vec{\nabla}\times\mathbf{E})]_1 = \frac{1}{2}\left[\frac{\partial}{\partial x_1}\left(E_1^2\right) + \frac{\partial}{\partial x_2}\left(E_1 E_2\right)\right.$$

$$\left. + \frac{\partial}{\partial x_3}\left(E_1 E_3\right)\right] - \frac{1}{2}\frac{\partial}{\partial x_1}\mathbf{E}^2\,, \tag{3.54}$$

and similarly for $\mathbf{B}$. After some trial and error, it appears that the tensor we are looking for is

$$T_{\alpha\beta} \equiv \frac{1}{4\pi}\left[E_\alpha E_\beta + B_\alpha B_\beta - \frac{1}{2}(\mathbf{E}\cdot\mathbf{E} + \mathbf{B}\cdot\mathbf{B})\,\delta_{\alpha\beta}\right]\,, \tag{3.55}$$

for then

$$\frac{d}{dt}\left(\mathbf{P}_{\text{mech}} + \int_V \mathbf{g}\, d^3x\right) = \int_V \vec{\nabla} \cdot \vec{\mathbf{T}}\, d^3x \,. \qquad (3.56)$$

This equation assumes the *dyadic* notation, wherein

$$(\vec{\nabla} \cdot \vec{\mathbf{T}})_\alpha = \sum_\beta \frac{\partial}{\partial x_\beta} T_{\alpha\beta} \,. \qquad (3.57)$$

Thus, using the divergence theorem,

$$\frac{d}{dt}\left(\mathbf{P}_{\text{mech}} + \int_V \mathbf{g}\, d^3x\right) = \oint_S \vec{\mathbf{T}} \cdot \hat{n}\, da \,. \qquad (3.58)$$

We interpret $\vec{\mathbf{T}} \cdot \hat{n}$ as the *momentum flux* normal to the boundary surface. That is, $\vec{\mathbf{T}} \cdot \hat{n}$ is the force per unit area transmitted across the surface S. Together with our success at establishing the dynamical properties of $\mathbf{E}$ and $\mathbf{B}$ in the previous sections, the nature of this Maxwell stress tensor[12] addresses the questions raised in the Introduction concerning the reality of the fields. We have now seen that they not only carry energy and momentum, but they also allow us to determine the electromagnetic force on a system through a consideration of $T_{\alpha\beta}$ written in terms of the local field components. In closing this section, we note that the volume in (3.58) is arbitrary, so an alternative representation of this equation is the equally useful differential form

$$\frac{\partial}{\partial t}\left(\mathbf{p}_{\text{mech}} + \mathbf{g}\right) = \vec{\nabla} \cdot \vec{\mathbf{T}} \,, \qquad (3.59)$$

where $\mathbf{p}_{\text{mech}}$ is the mechanical momentum density of the system and a partial derivative is now taken with respect to time since both $\mathbf{p}_{\text{mech}}$ and $\mathbf{g}$ may be functions of several variables.

3.3.4   Conservation of Angular Momentum

The derivation of the electromagnetic field angular momentum shares the same tactical approach as that of the linear momentum. Let us define the mechanical angular momentum density of our system as

$$\mathbf{l}_{\text{mech}} = \mathbf{r} \times \mathbf{p}_{\text{mech}} \,, \qquad (3.60)$$

---

[12] An extended discussion of the Maxwell stress tensor, taking into account the forces in fluids and solids, is given in Landau and Lifshitz (1975a).

where $\mathbf{p}_{mech}$ is the mechanical momentum density. Then,

$$\frac{\partial \mathbf{l}_{mech}}{\partial t} = \mathbf{r} \times \left(\rho \mathbf{E} + \frac{1}{c} \mathbf{J} \times \mathbf{B} \right) , \tag{3.61}$$

so that substitution for ρ and $\mathbf{J}$ from Maxwell's equations leads to

$$\frac{\partial}{\partial t} \left[\mathbf{l}_{mech} + \frac{1}{4\pi c} \mathbf{r} \times (\mathbf{E} \times \mathbf{B}) \right] =$$

$$\frac{1}{4\pi} \mathbf{r} \times [\mathbf{E}(\vec{\nabla} \cdot \mathbf{E}) - \mathbf{E} \times (\vec{\nabla} \times \mathbf{E}) + \mathbf{B}(\vec{\nabla} \cdot \mathbf{B}) - \mathbf{B} \times (\vec{\nabla} \times \mathbf{B})] . \tag{3.62}$$

Using our definition of the Maxwell stress tensor, we can simplify this equation considerably:

$$\frac{\partial}{\partial t} (\mathbf{l}_{mech} + \mathbf{l}_{em}) = \mathbf{r} \times \vec{\nabla} \cdot \vec{\mathbf{T}} , \tag{3.63}$$

where

$$\boxed{\mathbf{l}_{em} \equiv \mathbf{r} \times \mathbf{g}} \tag{3.64}$$

now has the simple interpretation of being the electromagnetic field angular momentum density. In integral form (with $\partial/\partial t \to d/dt$ now, since the integrated angular momenta are functions only of t),

$$\frac{d}{dt} \left(\mathbf{L}_{mech} + \int_V \mathbf{l}_{em} \, d^3x \right) = \int_S (\mathbf{r} \times \vec{\mathbf{T}}) \cdot \hat{n} \, da , \tag{3.65}$$

which follows from the fact that

$$\mathbf{r} \times \vec{\nabla} \cdot \vec{\mathbf{T}} = \vec{\nabla} \cdot (\mathbf{r} \times \vec{\mathbf{T}}) , \tag{3.66}$$

since $\vec{\nabla} \times \mathbf{r} = 0$. Not surprisingly, the right-hand side of this equation represents the integrated torque density due to the fields over the boundary surface S.

4

ELECTROMAGNETIC
WAVES AND RADIATION

What we have done for the electric and magnetic fields in Chapter 3 is equivalent to showing that particles have certain dynamical properties that are expressible in terms of the characteristics of the motion. For example, a particle's momentum $\mathbf{p} = m\mathbf{v}$ depends on its inertia and its velocity; so too do its kinetic energy and angular momentum. As such, a particle's dynamical attributes can change as $\mathbf{v}$ changes along its path. For particles, a complete description of the dynamics requires a solution for their trajectory, which provides $\mathbf{v}(t)$ and $\mathbf{r}(t)$ subject to the initial conditions. In contrast, when fluctuations of the fields (e.g., waves) are moving in vacuum, their speed is invariant (since all observers measure the same speed of light c). The field amplitudes, however, are not. Thus, for the fields, the momentum and angular momentum densities, and the energy flux change because the field intensities $\mathbf{E}$ and $\mathbf{B}$ vary in time and space. To complete the description of the field dynamics, we therefore need to know the temporal and spatial behavior of the field amplitudes (which also includes information on the wave vector, or propagation direction). This is the subject of the present chapter. We shall see that the fields may be either "attached" to the sources, or become completely detached in the radiation zone, and that their dynamical properties differ depending on which of these situations applies.

4.1   ELECTROMAGNETIC WAVES

In a region of space where there are no free sources, Maxwell's equations reduce to the simple form given in (1.75)–(1.78). We shall assume here that the medium is nonconducting, because otherwise Ohm's law $\mathbf{J} = \sigma\,\mathbf{E}$ results in a current density due to the presence of an electric field, which in turn acts as a source for additional fields. As we saw in § 1.3.3, this set of relations leads to a pair of wave equations for **E** and **B**. One of their solutions is a set of fields that display planar fluctuations:

$$\mathbf{E} = \mathbf{E}_0 \exp\{i(\mathbf{k} \cdot \mathbf{x} - \omega t)\} \,, \tag{4.1}$$

and

$$\mathbf{B} = \mathbf{B}_0 \exp\{i(\mathbf{k} \cdot \mathbf{x} - \omega t)\} \,, \tag{4.2}$$

where $\mathbf{E}_0$ and $\mathbf{B}_0$ are constant vectors, and

$$\mathbf{k} \cdot \mathbf{k} = \left(\frac{\omega}{v}\right)^2 \,, \tag{4.3}$$

with

$$v = \frac{c}{\sqrt{\mu\varepsilon}} \,. \tag{4.4}$$

We will consider the case of radial waves separately in § 8.2.

Let us now examine these solutions further and see what other properties are suggested by Maxwell's equations. The electric field must satisfy Gauss's law (1.75), which (with $\varepsilon = $ constant) leads to the condition

$$\vec{\nabla} \cdot \mathbf{E} = (ik_1 E_{01} + ik_2 E_{02} + ik_3 E_{03}) \exp\{i(\mathbf{k} \cdot \mathbf{x} - \omega t)\} = 0 \,. \tag{4.5}$$

That is,

$$\mathbf{k} \cdot \mathbf{E}_0 = 0 \,, \tag{4.6}$$

and it is trivial to show that **B** must satisfy the same constraint,

$$\mathbf{k} \cdot \mathbf{B}_0 = 0 \,. \tag{4.7}$$

This important result shows that electromagnetic waves in *nonconducting* media must be *transverse* to the propagation vector **k**.

Using the permutation tensor notation introduced in § 2.2.1, we see from the curl equations that

$$\varepsilon_{ijk} \, \hat{u}_i (\partial_j B_k - \partial_k B_j) = \frac{\mu\varepsilon}{c}(-i\omega)\mathbf{E} \,, \tag{4.8}$$

where $\hat{u}_i$ is the unit coordinate vector in the direction i. That is,

$$i\varepsilon_{ijk} \, \hat{u}_i (k_j B_k - k_k B_j) = \frac{\mu\varepsilon}{c}(-i\omega)\mathbf{E} \,, \tag{4.9}$$

and so

$$\mathbf{k} \times \mathbf{B} = -\frac{\mu\varepsilon}{c}\omega\,\mathbf{E} \,, \tag{4.10}$$

which may also be expressed as

$$\hat{k} \times \mathbf{B}_0 = -\sqrt{\mu\varepsilon}\,\mathbf{E}_0 \ . \tag{4.11}$$

Another way of writing this is to take the cross product of both sides with $\hat{k}$, which yields

$$\mathbf{B}_0 = \sqrt{\mu\varepsilon}\,\hat{k} \times \mathbf{E}_0 \tag{4.12}$$

when we use Equation (4.7). Thus, the combination of Equations (4.6), (4.7), and (4.11) requires that the vectors $\hat{k}$, $\mathbf{E}_0$, and $\mathbf{B}_0$ form an orthogonal set.

In general, $\mathbf{E}_0$ and $\mathbf{B}_0$ are complex, which alters slightly the manner in which we describe the dynamical characteristics of the field. For example, the Poynting vector (3.44), which gives the electromagnetic energy flux, would be "double counting" if we retained it in its present form. Instead, the Poynting vector must then be written as

$$\mathbf{S} = \frac{1}{2}\frac{c}{4\pi}\,\mathbf{E} \times \mathbf{H}^* \ , \tag{4.13}$$

where $\mathbf{E}$ and $\mathbf{H}$ are the measured fields at the point where $\mathbf{S}$ is evaluated. Note that we have made one additional modification in this expression to make it more general, writing $\mathbf{S}$ in terms of $\mathbf{E}$ and $\mathbf{H}$, rather than $\mathbf{B}$. This is because even though $\mathbf{B}$ is the applied induction, the actual field that carries the energy and momentum in media is $\mathbf{H}$. We discussed in § 1.2 how $\mathbf{H}$ represents the portion of the magnetic field produced by the external sources only, and that it excludes the magnetization $\mathbf{M}$, whose magnitude is dictated by the properties of the medium. Since the magnetization is "attached" to the medium, it constitutes a component of its internal mechanical structure, and so it cannot contribute to the energy flow due to the externally imposed fields. Particular attention should be paid to the factor of $1/2$ in this expression, which arises from the definition that the energy flux is given by the *real* part of $\mathbf{S}$:

$$\mathbf{S} = \frac{1}{2}\frac{c}{4\pi}\left[(\mathbf{E}_R \times \mathbf{H}_R) + (\mathbf{E}_I \times \mathbf{H}_I) + i(\mathbf{E}_I \times \mathbf{H}_R - \mathbf{E}_R \times \mathbf{H}_I)\right] \ . \tag{4.14}$$

When the field is harmonic, so that its time dependence is $e^{i\omega t}$, the time average of the square of a physical variable $\mathbf{A}$ pertaining to the field is obtained by first finding the real part of $\mathbf{A}$, squaring it, and then averaging the result over one period. Since $\mathbf{A}_R = (\mathbf{A} + \mathbf{A}^*)/2$, it is straightforward to see that the time average must be $\langle \mathbf{A}_R^2 \rangle = \mathbf{A} \cdot \mathbf{A}^*/2$. For plane waves, of course, the field amplitudes are $\mathbf{E}_0$ and $\mathbf{B}_0 = \sqrt{\mu\varepsilon}\,\hat{k} \times \mathbf{E}_0$ (see Equations [4.1] and [4.12]). Thus, with $\mathbf{H} = \mathbf{B}/\mu$, we get

$$\langle \mathbf{S}_R \rangle = \frac{c}{8\pi}\sqrt{\frac{\varepsilon}{\mu}}\,|\mathbf{E}_0|^2\,\hat{k} \ . \tag{4.15}$$

And since this energy flows with velocity $v = c/\sqrt{\mu\varepsilon}$, the corresponding time-averaged energy density is

$$u = \frac{|\langle \mathbf{S}_R \rangle|}{v} = \frac{\varepsilon}{8\pi} |\mathbf{E}_0|^2 . \tag{4.16}$$

Note that this is the *total* time-averaged energy density in the wave, not just the energy density associated with its electric field component, which would be just half of this value.

4.2 POLARIZATION AND STOKES PARAMETERS

Since $\mathbf{E}$, $\mathbf{B}$, and $\mathbf{k}$ form an orthogonal set, the most general way to write the electric field vector is

$$\mathbf{E}(\mathbf{x}, t) = (\mathbf{E}_{01} + \mathbf{E}_{02}) \exp\{i(\mathbf{k} \cdot \mathbf{x} - i\omega t)\} \equiv (\mathrm{E}_{01}\hat{\epsilon}_1 + \mathrm{E}_{02}\hat{\epsilon}_2) \exp\{i(\mathbf{k} \cdot \mathbf{x} - i\omega t)\} , \tag{4.17}$$

where $\hat{\epsilon}_1$, $\hat{\epsilon}_2$, and $\mathbf{k}$ constitute an alternative set of orthogonal vectors. But what is the benefit of this? After all, it would seem that we can just rotate the vectors $\hat{\epsilon}_1$ and $\hat{\epsilon}_2$ arbitrarily and we haven't learned anything. But this is only true when $\mathbf{E}_{01}$ and $\mathbf{E}_{02}$ are real or have the same phase. In general,

$$\mathrm{E}_{01} = |\mathrm{E}_{01}| \exp(i\phi_1) , \tag{4.18}$$

$$\mathrm{E}_{02} = |\mathrm{E}_{02}| \exp(i\phi_2) , \tag{4.19}$$

so that

$$\mathbf{E}(\mathbf{x}, t) = [\hat{\epsilon}_1 |\mathrm{E}_{01}| + \hat{\epsilon}_2 |\mathrm{E}_{02}| \exp\{i(\phi_2 - \phi_1)\}] \exp\{i\mathbf{k} \cdot \mathbf{x} - i\omega t + i\phi_1\} . \tag{4.20}$$

The overall phase of the field, ϕ_1, represents just a constant rotation in the complex plane and does not affect the physics. We can therefore drop it and effectively write the field as

$$\mathbf{E}(\mathbf{x}, t) = [\hat{\epsilon}_1 |\mathrm{E}_{01}| + \hat{\epsilon}_2 |\mathrm{E}_{02}| \exp\{i(\phi_2 - \phi_1)\}] \exp\{i\mathbf{k} \cdot \mathbf{x} - i\omega t\} . \tag{4.21}$$

To understand the behavior of the actual electric field, we need to consider the real part $\mathrm{Re}(\mathbf{E})$ of Equation (4.21). We see that when $\phi_2 - \phi_1 = 0$, the wave is *linearly polarized*, which is to say that $\mathrm{Re}(\mathrm{E}_1) = |\mathrm{E}_{01}| \cos(\mathbf{k} \cdot \mathbf{x} - \omega t)$ and $\mathrm{Re}(\mathrm{E}_2) = |\mathrm{E}_{02}| \cos(\mathbf{k} \cdot \mathbf{x} - \omega t)$, so that $\arctan[\mathrm{Re}(\mathrm{E}_2)/\mathrm{Re}(\mathrm{E}_1)]$ remains constant as the field evolves in space and time. That is, the field vector oscillates along a constant direction making an angle $\arctan[\mathrm{Re}(E_2)/\mathrm{Re}(E_1)]$ with respect to the $\hat{\epsilon}_1$ direction. However, when $\phi_2 - \phi_1 \neq 0$, the wave is instead *elliptically polarized* and the electric vector rotates around $\mathbf{k}$. The easiest way to see this is to note that because of the phase difference between $\mathbf{E}_{01}$ and $\mathbf{E}_{02}$, the field component in the $\hat{\epsilon}_1$ direction passes through its nodes at different spatial locations and/or times compared with the other one. They do not change proportionately, and the net effect of this is a rotation in the $\hat{\epsilon}_1 - \hat{\epsilon}_2$ plane. According to Equation (4.21), $\mathbf{E}$ sweeps around once every $2\pi/\omega$ seconds, and so the angular frequency of this rotation must be ω. It's easy to see this in the special case where $|E_{01}| = |E_{02}|$ with $\phi_2 - \phi_1 = \pm\pi/2$, for then the amplitude of the electric field $|\mathbf{E}|$ is constant, and $\mathrm{Re}(E_1) = |E_{01}|\cos(\mathbf{k} \cdot \mathbf{x} - \omega t)$ and $\mathrm{Re}(E_2) = \mp|E_{01}|\sin(\mathbf{k} \cdot \mathbf{x} - \omega t)$. Clearly, $\mathbf{E}$ here rotates around $\mathbf{k}$ with a constant magnitude and an angular frequency ω. This special case constitutes *circular* polarization (Figure 4.1).

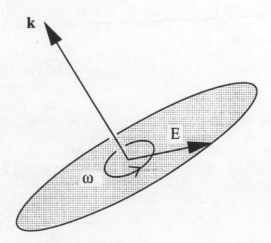

Figure 4.1 The polarization vector $\mathbf{E}$ is shown relative to the propagation direction $\mathbf{k}$. When $\phi_2 - \phi_1 \neq 0$, $\mathbf{E}$ rotates about its axis with an angular frequency ω, corresponding to the frequency of the wave. However, the magnitude of $\mathbf{E}$ is constant only when $\phi_2 - \phi_1 = \pm\pi/2$ and $|E_{01}| = |E_{02}|$, in which case the wave is circularly polarized.

An alternative general expression for $\mathbf{E}$ is in terms of the vectors

$$\hat{\epsilon}_\pm \equiv \frac{1}{\sqrt{2}}(\hat{\epsilon}_1 \pm i\hat{\epsilon}_2) . \tag{4.22}$$

In this instance,

$$\mathbf{E}(\mathbf{x}, t) = (\mathrm{E}_{0+}\hat{\epsilon}_+ + \mathrm{E}_{0-}\hat{\epsilon}_-) \exp\{i(\mathbf{k} \cdot \mathbf{x} - i\omega t)\} \,. \tag{4.23}$$

Following our discussion in the previous paragraph, it is apparent that $\hat{\epsilon}_+$ represents a positive helicity, i.e., the rotation of this field component is in the right-hand sense because the $\hat{\epsilon}_2$ piece leads the other one (by a phase exactly equal to $\pi/2$). For analogous reasons, $\hat{\epsilon}_-$ represents the negative helicity. We can see how this sense of rotation comes about more fundamentally by considering the two components E_1 and E_2 in $\mathbf{E} = \mathrm{E}_0(\hat{\epsilon}_1 + i\hat{\epsilon}_2) \exp\{i(\mathbf{k} \cdot \mathbf{x} - \omega t)\}$, which corresponds to an electric field with $\mathrm{E}_{0-} = 0$. Taking the real part of $\mathbf{E}$, we infer that

$$\mathrm{Re}(\mathrm{E}_1) \propto \cos(kz - \omega t) \,, \tag{4.24}$$

whereas

$$\mathrm{Re}(\mathrm{E}_2) \propto -\sin(kz - \omega t) \,. \tag{4.25}$$

Thus, for a fixed z, increasing t for an electric field vector in the fourth quadrant results in an increasing $\mathrm{Re}(\mathrm{E}_1)$ while $\mathrm{Re}(\mathrm{E}_2)$ approaches zero from negative values. That is, the electric field vector is rotating clockwise when seen from behind the $x - y$ plane in the direction of $\mathbf{k}$.

Having said this, the obvious next question is: how do we determine the polarization of a field vector? Clearly, we can describe the "internal" structure of the electric field with these two elegant basis vector systems, but unless we have a way of measuring the various components, this would not have much practical interest. Stokes parameters are quantities defined in terms of the projected amplitudes along each of the basis directions, such that together they allow us to isolate the various dependences of $\mathbf{E}$ on the phase and component amplitudes. In terms of the $(\hat{\epsilon}_1, \hat{\epsilon}_2)$ basis, the four Stokes parameters (based on the notation of Born and Wolf 1970) are

$$S_0 = |\hat{\epsilon}_1 \cdot \mathbf{E}|^2 + |\hat{\epsilon}_2 \cdot \mathbf{E}|^2 = |\mathrm{E}_{01}|^2 + |\mathrm{E}_{02}|^2 \,, \tag{4.26}$$

$$S_1 = |\hat{\epsilon}_1 \cdot \mathbf{E}|^2 - |\hat{\epsilon}_2 \cdot \mathbf{E}|^2 = |\mathrm{E}_{01}|^2 - |\mathrm{E}_{02}|^2 \,, \tag{4.27}$$

$$S_2 = 2\mathrm{Re}\left[(\hat{\epsilon}_1 \cdot \mathbf{E})^*(\hat{\epsilon}_2 \cdot \mathbf{E})\right] = 2|\mathrm{E}_{01}||\mathrm{E}_{02}|\cos(\phi_2 - \phi_1) \,, \tag{4.28}$$

$$S_3 = 2\mathrm{Im}\left[(\hat{\epsilon}_1 \cdot \mathbf{E})^*(\hat{\epsilon}_2 \cdot \mathbf{E})\right] = 2|\mathrm{E}_{01}||\mathrm{E}_{02}|\sin(\phi_2 - \phi_1) \,. \tag{4.29}$$

It is sometimes convenient to use an alternative definition of these four parameters in terms of the $(\hat{\epsilon}_+, \hat{\epsilon}_-)$ basis. Here, S_0 and S_1 contain information regarding the amplitudes of linear polarization, whereas S_2 and S_3 say something about the phases. Knowing these parameters (e.g., by passing a wave through perpendicular polarization filters) is sufficient for us to determine the amplitudes and relative phase of the field components.

4.3 REFLECTION AND REFRACTION

We learned in Chapter 2 how important surfaces and boundary conditions are in determining the electrostatic potential within a well-defined region of space. Time-dependent fields can also be bounded, but because of their dynamic nature, analyzing the field conditions near a surface can be much more complicated due to the various possible outcomes when a field is incident on the boundary. These include the partial reflection and partial refraction of the wave, which depend on a host of physical conditions, such as the type of material at the interface and the angle of incidence.

We do not wish here to divert greatly our attention away from the main theme of our development, so we shall consider only a straightforward situation with a high degree of symmetry and a simple geometry in order to learn the essential elements of handling these boundary conditions.[13] We will consider a plane electromagnetic wave, such as that described in §§ 4.1 and 4.2 above, incident on an interface between two media characterized by dielectric constants ε_1 and ε_2, and magnetic permeability μ_1 and μ_2, respectively. As we shall see, the solution for the field is found by adding the incident wave to the two other fields in the problem, viz. the reflected and transmitted (or refracted) waves.

To make the problem a little more specific, let us examine what happens when the incident wave is circularly polarized. Then, the electric field is

$$\mathbf{E}(\mathbf{x}, t) = \mathrm{E}_0(\hat{\epsilon}_1 + i\hat{\epsilon}_2) \exp\{i(\mathbf{k} \cdot \mathbf{x} - \omega t)\} , \qquad (4.30)$$

where E_0 is here taken to be real. In that case,

$$\mathrm{Re}(\mathbf{E}_1) = \mathrm{E}_0 \cos(\mathbf{k} \cdot \mathbf{x} - \omega t) , \qquad (4.31)$$

and

$$\mathrm{Re}(\mathbf{E}_2) = -\mathrm{E}_0 \sin(\mathbf{k} \cdot \mathbf{x} - \omega t) , \qquad (4.32)$$

so this represents a wave with positive helicity. For convenience, let us also take the vectors $\hat{\epsilon}_1$ and $\hat{\epsilon}_2$ to be such that $\hat{\epsilon}_1$ lies in the plane of incidence (with

---

[13]One of the best treatments of electromagnetic waves in media is that of Landau and Lifshitz (1975b).

respect to the interface between the two media). The fact that $\hat{\epsilon}_1$, $\hat{\epsilon}_2$, and $\mathbf{k}$ form an orthogonal set then automatically fixes the direction of $\hat{\epsilon}_2$. Throughout our discussion, it will be understood that $\mathbf{B} = \sqrt{\mu\varepsilon}\,\hat{\mathbf{k}} \times \mathbf{E}$, so if we know what happens to $\mathbf{E}$, we can immediately determine the fate of the magnetic field as well. Our approach will be to consider what happens to the two incident field components $\mathbf{E}_{i1} = \mathrm{E}_0\,\hat{\epsilon}_1$ and $\mathbf{E}_{i2} = \mathrm{E}_0\,i\hat{\epsilon}_2$ separately (Figure 4.2).

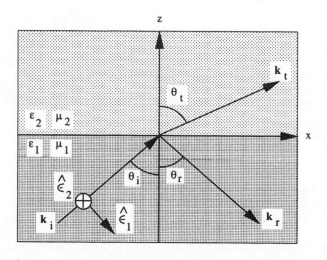

Figure 4.2 The boundary value problem for a wave incident on the interface from a medium with ε_1 and μ_1 into an adjoining medium with ε_2 and μ_2. The basis vectors $\hat{\epsilon}_1$ and $\hat{\epsilon}_2$ are oriented such that $\hat{\epsilon}_1$ is in the plane of the paper and $\hat{\epsilon}_2$ points into the page.

For each component of polarization, we designate the field incident from medium 1 onto the interface to be $\mathbf{E}_i$, that reflected back into medium 1 to be $\mathbf{E}_r$, and that transmitted across the interface into medium 2 to be $\mathbf{E}_t$. Their corresponding wave vectors are, respectively, $\mathbf{k}_i$, $\mathbf{k}_r$, and $\mathbf{k}_t$. Thus, we have

$$\mathbf{E}_i = \mathbf{A}\,\exp\{i(\mathbf{k}_i \cdot \mathbf{x} - \omega t)\}\,, \tag{4.33}$$

$$\mathbf{E}_r = \mathbf{R}\,\exp\{i(\mathbf{k}_r \cdot \mathbf{x} - \omega t)\}\,, \tag{4.34}$$

$$\mathbf{E}_t = \mathbf{T}\,\exp\{i(\mathbf{k}_t \cdot \mathbf{x} - \omega t)\}\,. \tag{4.35}$$

We cannot assume any particular direction for $\mathbf{k}_r$ and $\mathbf{k}_t$ a priori. However, we do know that

$$|\mathbf{k}_i| = |\mathbf{k}_r| = \sqrt{\mu_1 \varepsilon_1} \, \frac{\omega}{c} \, , \tag{4.36}$$

and

$$|\mathbf{k}_t| = \sqrt{\mu_2 \varepsilon_2} \, \frac{\omega}{c} \, . \tag{4.37}$$

Now, we also realize from Equation (2.120) at the boundary that the perpendicular component of $\mathbf{B}$ must be continuous at $z = 0$. But in order for this to be the case for all $\mathbf{x}$, we need

$$(\mathbf{k}_i \cdot \mathbf{x})|_{z=0} = (\mathbf{k}_r \cdot \mathbf{x})|_{z=0} = (\mathbf{k}_t \cdot \mathbf{x})|_{z=0} \, . \tag{4.38}$$

Thus, for the way we've chosen our axes,

$$k_{ix} \, x = k_{rx} \, x + k_{ry} \, y \, . \tag{4.39}$$

But this must be true for all y as well, and so $k_{ry} = 0$, which implies

$$k_{ix} = k_{rx} \, . \tag{4.40}$$

Similarly, from the second equality in (4.38), we get

$$k_{ty} = 0 \, . \tag{4.41}$$

The net result is that all three vectors $\mathbf{k}_r$, $\mathbf{k}_i$, and $\mathbf{k}_t$ are *coplanar*.

We see immediately that because $|\mathbf{k}_i| = |\mathbf{k}_r|$, Equation (4.40) gives

$$\sin \theta_i = \sin \theta_r \, , \tag{4.42}$$

that is, the angle of incidence equals the angle of reflection. Second, Equation (4.36) implies that

$$|\mathbf{k}_i| \sin \theta_i = |\mathbf{k}_t| \sin \theta_t \, , \tag{4.43}$$

so that

$$\frac{\sin \theta_1}{\sin \theta_2} = \sqrt{\frac{\mu_2 \varepsilon_2}{\mu_1 \varepsilon_1}} \, , \tag{4.44}$$

which is *Snell's law* of refraction.

In the absence of a surface charge, the tangential component of $\mathbf{E}$ along the interface must be continuous, so

$$\mathbf{A}_2 + \mathbf{R}_2 = \mathbf{T}_2 \, , \tag{4.45}$$

where subscript 2 refers to the fields parallel to the basis vector $\hat{\epsilon}_2$. All three of these vectors point in the same direction, so Equation (4.45) is more conveniently written simply as an algebraic equation:

$$A_2 + R_2 = T_2 \ . \tag{4.46}$$

We get additional constraints using the known form of the magnetic field $\mathbf{B} = \sqrt{\mu \varepsilon}\, \hat{k} \times \mathbf{E}$, which gives

$$
\begin{aligned}
B_{ilx} &= -\sqrt{\mu_1 \varepsilon_1}\, A_2 \cos \theta_i \ , \\
B_{rlx} &= \sqrt{\mu_1 \varepsilon_1}\, R_2 \cos \theta_r \ , \\
B_{tlx} &= -\sqrt{\mu_2 \varepsilon_2}\, T_2 \cos \theta_t \ .
\end{aligned}
\tag{4.47}
$$

Requiring continuity of these tangential components of $\mathbf{B}$ at $z = 0$ gives a second equation for the $\hat{\epsilon}_2$ components of $\mathbf{E}$:

$$B_{ilx} + B_{rlx} = B_{tlx}$$

or

$$\sqrt{\mu_1 \varepsilon_1}(A_2 - R_2) \cos \theta_i = \sqrt{\mu_2 \varepsilon_2}\, T_2 \cos \theta_t \ . \tag{4.48}$$

It is straightforward to solve the pair of simultaneous equations (4.46) and (4.48) with the aid of Snell's law (4.44), and the result is

$$\frac{T_2}{A_2} = \frac{2 \cos \theta_i \sin \theta_t}{\sin(\theta_i + \theta_t)} \ , \tag{4.49}$$

and

$$\frac{R_2}{A_2} = -\frac{\sin(\theta_i - \theta_t)}{\sin(\theta_i + \theta_t)} \ . \tag{4.50}$$

For the other component of $\mathbf{E}$ (in the $\hat{\epsilon}_1$ direction), we have

$$\cos \theta_i (A_1 - R_1) = \cos \theta_t\, T_1 \ , \tag{4.51}$$

and continuity of $\mathbf{B}_2$ results in a second equation

$$\sqrt{\mu_1 \varepsilon_1}\, (A_1 + R_1) = \sqrt{\mu_2 \varepsilon_2}\, T_1 \ . \tag{4.52}$$

Again eliminating the unknowns and using Snell's law results in

$$\frac{R_1}{A_1} = \frac{\tan(\theta_i - \theta_t)}{\tan(\theta_i + \theta_t)} \ , \tag{4.53}$$

and

$$\frac{T_1}{A_1} = \frac{2\cos\theta_i \sin\theta_t}{\sin(\theta_i + \theta_t)\cos(\theta_i - \theta_t)} \ , \tag{4.54}$$

and this, in principle, solves the problem.

Example 4.1. Let us consider what we need in order to design an interface that transforms circularly polarized electromagnetic waves into elliptically polarized waves whose ratio of major to semimajor axes is $\alpha = 1.3$. In this problem, we need

$$\frac{T_1}{T_2} = \alpha = \frac{\sin(\theta_i + \theta_t)}{\sin(\theta_i + \theta_t)\cos(\theta_i - \theta_t)} \ . \tag{4.55}$$

Suppose the wave is incident from air ($\mu_1\varepsilon_1 \approx 1$) into a glass for which $\sqrt{\mu_2\varepsilon_2} \approx 1.5$. Then, from Snell's law,

$$\sin\theta_t = 0.667 \sin\theta_i \ , \tag{4.56}$$

and so $\alpha \approx 1.3$ when the circularly polarized wave is incident at an angle $i = 75°$.

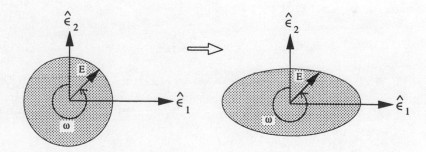

Figure 4.3 The conversion of an initially circularly polarized wave into an elliptically polarized wave. The ratio of the semimajor to semiminor transmitted amplitudes for this example is 1.3.

Is it obvious why T_1 should be larger than T_2? Let's see what happens to $\mathbf{E}_1$. Since it points roughly in the direction of $\mathbf{k}_r$, this component is attenuated more in reflection than $\mathbf{E}_2$ which is perpendicular to $\mathbf{k}_r$. That is, the reflected wave is expected to be polarized more in the $\hat{\epsilon}_2$ direction than the incident wave. Thus, T_1 gains at R_1's expense (Figure 4.3).

4.4 TIME HARMONIC FIELDS IN MATTER

All matter contains charged particles. Thus, an electromagnetic wave incident on a sample of matter sets these charges in motion and in effect produces a spatial distribution of currents. The different nature of the currents from material to material results in a variation of macroscopic properties, such as conduction and insulation. For most of our study thus far, we have assumed a dielectric constant ε independent of frequency ω. But because of these induced currents, this assumption is at best only an approximation, as we shall see from the following simple model for $\varepsilon(\omega)$.

We have written the effects of polarizability of the medium in terms of a polarization vector $\mathbf{P}$ (i.e., the dipole moment per unit volume):

$$\mathbf{D} = \mathbf{E} + 4\pi\,\mathbf{P} = \varepsilon\,\mathbf{E}\,. \tag{4.57}$$

In a low-density medium, one in which the local field is not very different from the applied field, the equation of motion for an electron bound to a harmonic restoring force

$$\mathbf{F} = -m_e\,{\omega_0}^2\,\mathbf{x}\,, \tag{4.58}$$

and acted on by an electric field $\mathbf{E}(\mathbf{x},t)$ is

$$m_e\left(\frac{d^2\mathbf{x}}{dt^2} + \gamma\frac{d\mathbf{x}}{dt} + {\omega_0}^2\mathbf{x}\right) = e\,\mathbf{E}(\mathbf{x},t)\,. \tag{4.59}$$

Here, ω_0 is the restoring frequency, m_e is the electronic mass, e is its charge, and γ measures the (phenomenological) damping force. We ignore magnetic forces under the assumption that $v/c \ll 1$. Thus, if $\mathbf{E}$ is harmonic, say $\mathbf{E} = \mathbf{E}_\omega \exp(-i\omega t)$, then

$$\mathbf{x} = \frac{e\,\mathbf{E}/m_e}{({\omega_0}^2 - \omega^2 - i\omega\gamma)}\,. \tag{4.60}$$

But the single charge dipole moment is,

$$\mathbf{p} = e\,\mathbf{x} = \frac{e^2\,\mathbf{E}/m_e}{({\omega_0}^2 - \omega^2 - i\omega\gamma)}\,, \tag{4.61}$$

so that if there are N scatterers per unit volume,

$$\mathbf{P} = N\,\mathbf{p} = \frac{e^2\,\mathbf{E}\,N/m_e}{({\omega_0}^2 - \omega^2 - i\omega\gamma)}\,. \tag{4.62}$$

According to the definition of ε in Equation (4.57), this leads to the result

$$\varepsilon(\omega) = 1 + \frac{4\pi N e^2}{m_e} \frac{1}{(\omega_0{}^2 - \omega^2 - i\omega\gamma)} . \qquad (4.63)$$

In reality, of course, a medium has a mixture of restoring frequencies so that $\varepsilon(\omega)$ looks rather more complicated than this, but the basic physical principle is the same.

This does not yet tells us much about how the fields behave once they enter the medium. To learn more about this aspect, we need to go back to Maxwell's equations and actually solve them for the propagation vector $\mathbf{k}$. But let us first be clear on the assumptions we are making here. Although we are obviously allowing the charges to move around so that they can create currents, we're going to insist that the medium remains neutral and so $\rho = 0$. Second, the various parameters characterizing the medium, such as ε and now $\mathbf{k}$, are allowed to vary with ω, but we will still retain the simplifying condition that they do not depend on space and time. We can therefore take Equations (1.14), (1.22), (1.47), and (1.49) and write them in the form

$$\vec{\nabla} \cdot \mathbf{E} = 0 , \qquad (4.64)$$

$$\vec{\nabla} \times \mathbf{E} + \frac{1}{c} \frac{\partial \mathbf{B}}{\partial t} = 0 , \qquad (4.65)$$

$$\vec{\nabla} \times \mathbf{B} - \frac{\mu\varepsilon}{c} \frac{\partial \mathbf{E}}{\partial t} = \frac{4\pi\mu}{c} \mathbf{J} , \qquad (4.66)$$

and

$$\vec{\nabla} \cdot \mathbf{B} = 0 . \qquad (4.67)$$

The usual approach in solving equations such as these is to separate out the dependence on $\mathbf{E}$ and $\mathbf{B}$, which we can do here by taking the curl of Equations (4.65) and (4.66), and substituting for $\vec{\nabla} \times \mathbf{E}$ and $\vec{\nabla} \times \mathbf{B}$. The result is two independent wave equations for the fields:

$$-\vec{\nabla}^2 \mathbf{E} + \frac{\mu\varepsilon}{c^2} \frac{\partial^2 \mathbf{E}}{\partial t^2} + \frac{4\pi\mu}{c^2} \frac{\partial \mathbf{J}}{\partial t} = 0 , \qquad (4.68)$$

and

$$-\vec{\nabla}^2 \mathbf{B} + \frac{\mu\varepsilon}{c^2} \frac{\partial^2 \mathbf{B}}{\partial t^2} - \frac{4\pi\mu}{c} \vec{\nabla} \times \mathbf{J} = 0 . \qquad (4.69)$$

Continuing with our assumption of harmonic fields and currents, we write

$$\mathbf{E} = \mathbf{E}_\omega \exp(-i\omega t) , \qquad (4.70)$$

$$\mathbf{B} = \mathbf{B}_\omega \exp(-i\omega t) , \tag{4.71}$$

$$\mathbf{J} = \mathbf{J}_\omega \exp(-i\omega t) , \tag{4.72}$$

for which the field wave equations simplify to

$$\left[\vec{\nabla}^2 + k^2(\omega)\right] \mathbf{E}_\omega = -\frac{i4\pi}{c} k(\omega)\sqrt{\frac{\mu}{\varepsilon}}\,\mathbf{J}_\omega , \tag{4.73}$$

and

$$\left[\vec{\nabla}^2 + k^2(\omega)\right] \mathbf{B}_\omega = -\frac{4\pi\mu}{c} \vec{\nabla} \times \mathbf{J}_\omega , \tag{4.74}$$

where

$$k(\omega) \equiv \frac{n(\omega)\,\omega}{c} , \tag{4.75}$$

and the so-called *index of refraction* is

$$n(\omega) \equiv \sqrt{\mu\varepsilon} . \tag{4.76}$$

Now we make the basic assumption—a generalized Ohm's law if you will—that

$$\mathbf{J}_\omega = \Gamma\,\mathbf{E}_\omega , \tag{4.77}$$

where Γ is an arbitrary (complex) parameter, which is often also a function of ω. This eliminates $\mathbf{J}_\omega$ from the equations, which now reduce to

$$\left[\vec{\nabla}^2 + k^2(\omega) + \frac{i4\pi}{c}\frac{k(\omega)n(\omega)\,\Gamma}{\varepsilon}\right] \mathbf{E}_\omega = 0 , \tag{4.78}$$

and

$$\left[\vec{\nabla}^2 + k^2(\omega) + \frac{i4\pi}{c}\frac{k(\omega)n(\omega)\Gamma}{\varepsilon}\right] \mathbf{B}_\omega = 0 . \tag{4.79}$$

The last step follows from using Faraday's law to eliminate $\vec{\nabla} \times \mathbf{E}_\omega$ once we substitute for $\mathbf{J}_\omega$ in Equation (4.74) using Ohm's law. Evidently, the fields propagate through this medium with a wave vector of magnitude k', where

$$k'^2(\omega) \equiv k^2(\omega) + \frac{i4\pi k(\omega)n(\omega)\Gamma}{c\varepsilon} , \tag{4.80}$$

since by analogy with the solution to Equations (1.84) and (1.87), we here infer that

$$\mathbf{E}_\omega = \mathbf{E}_0 \exp\{i\mathbf{k}'(\omega) \cdot \mathbf{x}\} = \mathbf{E}_0 \exp(i\mathbf{k}'_r \cdot \mathbf{x}) \exp(-\mathbf{k}'_i \cdot \mathbf{x}) , \tag{4.81}$$

and

$$\mathbf{B}_\omega = \mathbf{B}_0 \exp\{i\mathbf{k}'(\omega) \cdot \mathbf{x}\} = \mathbf{B}_0 \exp(i\mathbf{k}'_r \cdot \mathbf{x}) \exp(-\mathbf{k}'_i \cdot \mathbf{x}) . \tag{4.82}$$

The subscripts r and i here have their obvious meaning, denoting the real and imaginary parts of $\mathbf{k}'$, respectively. The corresponding complex index of refraction $\bar{n}$ is defined to be

$$\bar{n}^2 \equiv \left[\frac{k'(\omega)}{k(\omega)}\right]^2 = 1 + \frac{i4\pi n(\omega)\Gamma}{c\varepsilon k(\omega)} . \tag{4.83}$$

Let us now use what we have learned so far to see qualitatively how the behavior of the fields depends on the characteristics of the medium. In our simple model of the dielectric constant, we said that a typical charge's position is given by Equation (4.60). Thus, since $\mathbf{v}_\omega = -i\omega\,\mathbf{x}_\omega$ and $\mathbf{J}_\omega = Ne\mathbf{v}_\omega$, we infer from (4.77) that within the framework of our schematic approach

$$\Gamma = \frac{-iN\omega e^2}{m_e(\omega_0{}^2 - \omega^2 - i\omega\gamma)} . \tag{4.84}$$

We see that when Γ has a *real* part (it may or may not also have an imaginary part), $\mathbf{k}'$ has both real (k_r') and imaginary (k_i') components and therefore according to Equations (4.81) and (4.82), k_r' represents a propagation of the field whereas k_i' accounts for its attenuation with distance. Media with this property are conductors, and we shall discuss them further below. However, when Γ is purely imaginary, $\mathbf{k}'$ is real, and these media therefore have a real index of refraction. The fields propagate through these so-called dielectrics without any attenuation.

Example 4.2.
It is very instructive to consider special situations, such as the one in which $\omega_0 = 0$ and $\gamma = 0$. The first condition suggests immediately that the charges are not bound, and this type of matter is often referred to as a *plasma*. Looking at Equation (4.80) with the known form of Γ, we therefore learn that in this case

$$k'^2 = k^2 \left(1 - \frac{\omega_p^2}{\omega^2}\right) , \tag{4.85}$$

where

$$\omega_p \equiv \frac{4\pi Ne^2}{m_e\varepsilon} \tag{4.86}$$

is known as the *plasma frequency*. A plasma distinctively dichotomizes the incoming waves according to whether or not their frequency exceeds ω_p. The wavenumber k' is real for high-frequency fields, which therefore propagate through the medium, but is imaginary when $\omega < \omega_p$. The low-frequency waves are therefore damped within the plasma. But what does this mean, given that we explicitly chose to put $\gamma = 0$, which presumably implies that there is no dissipation? Clearly, the energy is taken out of the fields as they penetrate into the plasma, but instead of being dissipated as heat, the energy is apparently transformed into bulk oscillations of the gas. Although we cannot describe this process in detail here, our intuition tells us that when ω is very high, there is insufficient time for the gas to adjust to the field and to thereby permit the excitation of the oscillations. Lower frequency waves, on the other hand, continuously "pump" these modes, which therefore grow in energy and eventually attenuate the field.

Example 4.3. Suppose that the charges are unbound (i.e., $\omega_0 = 0$) but $\gamma \neq 0$. Then,

$$\Gamma = \frac{i\omega N e^2}{m_e(\omega^2 + i\omega\gamma)} \, , \tag{4.87}$$

which according to our discussion above tells us that we are dealing with a conductor. Experimentally, it is found for these materials that $\gamma \approx 10^{17}$ s^{-1}, and so for waves with $\omega \ll \gamma$ (i.e., for electromagnetic waves longward of the optical/UV portion of the spectrum), $\Gamma \approx Ne^2/m_e\gamma$ and its real component is much larger in magnitude than that of its imaginary piece. We note that in this situation, Γ is then normally called the *conductivity* σ:

$$\mathbf{J}_\omega = \sigma \mathbf{E}_\omega \equiv \frac{Ne^2}{m_e\gamma} \mathbf{E}_\omega \, . \tag{4.88}$$

In terms of σ,

$$k'^2 = \left[\frac{\omega n(\omega)}{c}\right]^2 + \frac{i4\pi n^2(\omega)\omega\sigma}{c^2\varepsilon} \, , \tag{4.89}$$

and when the second term is dominant (which is true for all wavelengths down to the infrared range),

$$k' \approx \sqrt{\frac{i4\pi n^2(\omega)\omega\sigma}{c^2\varepsilon}} = \frac{1+i}{\delta} \, , \tag{4.90}$$

where for obvious reasons,

$$\delta \equiv c\sqrt{\frac{\varepsilon}{2\pi\sigma\omega n^2}} \tag{4.91}$$

is known as the *skin depth* of the medium. Its importance is underscored by the application of this result to Equations (4.81) and (4.82), which show that the attenuation factor is here $\sim \exp(-x/\delta)$ so that the waves drop to $1/e$ of their external value within a depth δ of the surface. Apparently, when an electromagnetic field is incident on a conductor, most of its effect and the current it induces are confined to a thin skin near the conductor's surface.

4.5   WAVE GUIDES

So far in this chapter we have looked at the general properties of electromagnetic waves implied by Maxwell's equations, and we have investigated how these waves behave at a simple boundary where both reflection and refraction may occur, and we have studied the effects of the medium on the propagation of the fields. Before moving on to explore the properties of these waves as they become detached from their sources and turn into true radiation fields, we will complete this overview by considering the complementary situation in which the medium is a vacuum, but is, however, enclosed by a well-defined boundary such that the surface conditions explicitly determine the allowed field structure in this space. This situation is not unlike that of the long and thorough investigation we carried out in § 2.1, except that now the fields are time dependent and so they represent a transfer of energy and influence within the enclosed region—the wave guide.

Again, we consider sourceless fields because the medium is a vacuum. We shall see later on that a very useful boundary condition is provided when the enclosure is a perfect conductor. Thus, we are led once again to consider harmonic fields, which in this type of situation can greatly simplify the equations and lead to familiar forms with well-studied solutions. Assuming that the fields look like those specified in Equations (4.70) and (4.71), the sourceless Maxwell's equations within a nondissipative medium (characterized by a constant dielectric and permeability) are simply

$$\vec{\nabla} \cdot \mathbf{E}_\omega = 0 \,, \tag{4.92}$$

$$\vec{\nabla} \times \mathbf{E}_\omega = \frac{i\omega}{c} \mathbf{B}_\omega \,, \tag{4.93}$$

$$\vec{\nabla} \cdot \mathbf{B}_\omega = 0 \,, \tag{4.94}$$

and

$$\vec{\nabla} \times \mathbf{B}_\omega = -\frac{i\mu\varepsilon\omega}{c} \mathbf{E}_\omega \,. \tag{4.95}$$

As we have seen before, these expressions reduce to the wave equation for both $\mathbf{E}_\omega$ and $\mathbf{B}_\omega$:

$$\left(\vec{\nabla}^2 + \frac{\mu\varepsilon\omega^2}{c^2} \right) \mathbf{E}_\omega = 0 \,, \tag{4.96}$$

$$\left(\vec{\nabla}^2 + \frac{\mu\varepsilon\omega^2}{c^2} \right) \mathbf{B}_\omega = 0 \,. \tag{4.97}$$

For simplicity, we will assume that the medium is symmetric in the z-direction and define axes accordingly (Figure 4.4). We assume that the spatial dependence of the fields along this axis can be separated out from the transverse, so that

$$\mathbf{E}_\omega(\mathbf{x}) = \mathbf{E}_{\omega t}(x, y) \exp(ik_g z) , \tag{4.98}$$

$$\mathbf{B}_\omega(\mathbf{x}) = \mathbf{B}_{\omega t}(x, y) \exp(ik_g z) . \tag{4.99}$$

The Maxwell equations then turn into equations just for the transverse fields $\mathbf{E}_{\omega t}(x, y)$ and $\mathbf{B}_{\omega t}(x, y)$:

$$\left(\frac{\partial^2}{\partial x^2} + \frac{\partial^2}{\partial y^2} + \frac{\mu \varepsilon \omega^2}{c^2} - k_g^2 \right) \mathbf{E}_{\omega t}(x, y) = 0 , \tag{4.100}$$

$$\left(\frac{\partial^2}{\partial x^2} + \frac{\partial^2}{\partial y^2} + \frac{\mu \varepsilon \omega^2}{c^2} - k_g^2 \right) \mathbf{B}_{\omega t}(x, y) = 0 . \tag{4.101}$$

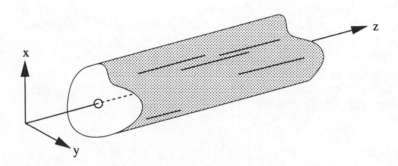

Figure 4.4 Segment of a wave guide whose symmetry axis is aligned with $\hat{z}$.

We will return to these in a moment. Now let's see what Maxwell's equations say about the relationships between the individual field components. The vector equation $\vec{\nabla} \times \mathbf{E}_\omega = i\omega \mathbf{B}_\omega / c$ yields three separate algebraic constraints, one for each of the three independent coordinate directions:

$$\frac{\partial E_{\omega t z}}{\partial y} - ik_g E_{\omega t y} = \frac{i\omega}{c} B_{\omega t x} , \tag{4.102}$$

$$-\frac{\partial E_{\omega t z}}{\partial x} + ik_g E_{\omega t x} = \frac{i\omega}{c} B_{\omega t y} , \tag{4.103}$$

$$\frac{\partial E_{\omega t y}}{\partial x} - \frac{\partial E_{\omega t x}}{\partial y} = \frac{i\omega}{c} B_{\omega t z} . \tag{4.104}$$

In similar fashion, the other curl equation $\vec{\nabla} \times \mathbf{B}_\omega = -i\omega\mu\varepsilon\mathbf{E}_\omega/c$ gives three more relationships:

$$\frac{\partial B_{\omega t z}}{\partial y} - ik_g B_{\omega t y} = -\frac{i\omega\mu\varepsilon}{c} E_{\omega t x} , \qquad (4.105)$$

$$-\frac{\partial B_{\omega t z}}{\partial x} + ik_g B_{\omega t x} = -\frac{i\omega\mu\varepsilon}{c} E_{\omega t y} , \qquad (4.106)$$

$$\frac{\partial B_{\omega t y}}{\partial x} - \frac{\partial B_{\omega t x}}{\partial y} = -\frac{i\omega\mu\varepsilon}{c} E_{\omega t z} . \qquad (4.107)$$

The equations available at our disposal for solving this coupled system suggest that we determine the x and y components of $\mathbf{E}$ and $\mathbf{B}$ in terms of $E_{\omega t z}$ and $B_{\omega t z}$ and then use Equations (4.96) and (4.97) to obtain the latter. For example, eliminating $E_{\omega t y}$ from (4.102) and (4.106) gives

$$B_{\omega t x} = -\frac{i}{(\omega/c)^2 - k_g^2} \left[\frac{\omega}{c} \frac{\partial E_{\omega t z}}{\partial y} - k_g \frac{\partial B_{\omega t z}}{\partial x} \right] . \qquad (4.108)$$

Similarly,

$$B_{\omega t y} = \frac{i}{(\omega/c)^2 - k_g^2} \left[\frac{\omega}{c} \frac{\partial E_{\omega t z}}{\partial x} + k_g \frac{\partial B_{\omega t z}}{\partial y} \right] , \qquad (4.109)$$

$$E_{\omega t x} = \frac{i}{(\omega/c)^2 - k_g^2} \left[\frac{\omega}{c} \frac{\partial B_{\omega t z}}{\partial y} + k_g \frac{\partial E_{\omega t z}}{\partial x} \right] , \qquad (4.110)$$

$$E_{\omega t y} = -\frac{i}{(\omega/c)^2 - k_g^2} \left[\frac{\omega}{c} \frac{\partial B_{\omega t z}}{\partial x} - k_g \frac{\partial E_{\omega t z}}{\partial y} \right] . \qquad (4.111)$$

In general, the solution will be a superposition of two configurations:

$$\text{transverse electric (TE) mode:} \qquad E_{\omega t z} = 0$$
$$\text{transverse magnetic (TM) mode:} \qquad B_{\omega t z} = 0 .$$

The motivation for this becomes clear in the case where the boundary of the medium is a conductor. Faraday's law (4.93) applied to the rectangular loop shown in Figure 4.5 gives

$$\int_S \vec{\nabla} \times \mathbf{E}_\omega \cdot \hat{n} \, da = \oint \mathbf{E}_\omega \cdot d\mathbf{1} = \frac{i\omega}{c} \int_S \mathbf{B} \cdot \hat{n} \, da , \qquad (4.112)$$

which leads to

$$\hat{k} \cdot (\mathbf{E}_{\omega t,1} - \mathbf{E}_{\omega t,2}) \, \Delta l \approx |\mathbf{B}_{\omega t}| \, \Delta l \, \Delta h \, \frac{\omega}{c} . \qquad (4.113)$$

We have assumed here that the contribution to the loop integral on the left-hand side from the Δh elements is negligible as $\Delta h \to 0$. Thus, if $\mathbf{B}_\omega$ is not singular, we get the straightforward boundary condition

$$\hat{k} \cdot \mathbf{E}_{\omega t,1} = \hat{k} \cdot \mathbf{E}_{\omega t,2} \,, \tag{4.114}$$

as $\Delta h \to 0$. In particular, if side 1 is a conductor, then the fact that $\mathbf{E}_{\omega t,1} = 0$ forces the condition $\hat{k} \cdot \mathbf{E}_{\omega t,2} = 0$ on the boundary. One possible solution therefore has $E_{\omega t z} = 0$ everywhere inside the medium, and this would correspond to a TE mode. A similar derivation using the other curl equation gives us the condition

$$\hat{k} \cdot \mathbf{B}_{\omega t,1} = \hat{k} \cdot \mathbf{B}_{\omega t,2} \,, \tag{4.115}$$

and so if the currents are zero, a complementary situation to the one we have just described is that in which $B_{\omega t z}$ is instead zero. This solution is the TM mode. The specification of the fields is completed when $E_{\omega t z}$ and $B_{\omega t z}$ are determined from the vacuum equations

$$\left(\frac{\partial^2}{\partial x^2} + \frac{\partial^2}{\partial y^2} + \frac{\omega^2}{c^2} - k_g^2 \right) E_{\omega t z} = 0 \,, \tag{4.116}$$

$$\left(\frac{\partial^2}{\partial x^2} + \frac{\partial^2}{\partial y^2} + \frac{\omega^2}{c^2} - k_g^2 \right) B_{\omega t z} = 0 \,. \tag{4.117}$$

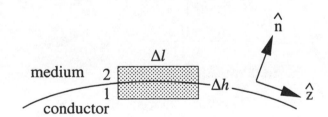

Figure 4.5 Evaluation of the boundary conditions at the interface between the internal wave guide medium and the conducting surface. The unit normal is $\hat{n}$, whereas the unit vector parallel to the surface is $\hat{z}$.

Below, we shall solve a straightforward, though very informative, wave guide problem that demonstrates all the techniques we have outlined above.[14]

[14]This subject has quite an extensive literature, especially in engineering physics. Three of the worthwhile references are Panofsky and Phillips (1962), Ramo, Whinnery, and Van Duzer (1965), and Borgnis and Papas (1968).

Example 4.4. Let us consider a conducting rectangular wave guide with side lengths a and b, enclosing a vacuum carrying an electromagnetic wave (Figure 4.6). For specificity, assume that the wave is composed solely of TM modes, for which $B_{\omega tz} = 0$. The solution to (4.116) contains terms like $\sin(k_x x)$, $\cos(k_x x)$, $\sin(k_y y)$, and $\cos(k_y y)$. However, since $E_{\omega tz}$ must be zero at the (conducting) boundaries, the only viable functional form is

$$E_{\omega tz} = E_0 \sin(k_x x) \sin(k_y y) \,, \tag{4.118}$$

where

$$k_x = \frac{m\pi}{a} \qquad m = 1, 2, \ldots, \tag{4.119}$$

and

$$k_y = \frac{n\pi}{b} \qquad n = 1, 2, \ldots. \tag{4.120}$$

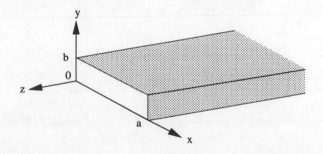

Figure 4.6 Rectangular wave guide with dimensions a and b pointing in the $\hat{z}$ direction. It is assumed for simplicity that the surface is a conductor and that the interior is a vacuum.

Substituting the *Ansatz* for $E_{\omega tz}$ into Equation (4.116) then shows that the wavenumber components must satisfy the constraint

$$k_g^2 = \left(\frac{\omega}{c}\right)^2 - k_x^2 - k_y^2 = \left(\frac{\omega}{c}\right)^2 - \pi^2 \left(\frac{m^2}{a^2} + \frac{n^2}{b^2}\right) \,. \tag{4.121}$$

Thus, with $B_{\omega tz} = 0$ for all x and y, the components of the fields defined in Equations (4.98) and (4.99) are easily seen to be

$$E_{\omega x} = \frac{iE_0 k_g k_x}{k^2 - k_g^2} \cos(k_x x) \sin(k_y y) \exp(ik_g z) \,, \tag{4.122}$$

$$E_{\omega y} = \frac{iE_0 k_g k_y}{k^2 - k_g^2} \sin(k_x x) \cos(k_y y) \exp(ik_g z) \,, \qquad (4.123)$$

$$E_{\omega z} = E_0 \sin(k_x x) \sin(k_y y) \exp(ik_g z) \,, \qquad (4.124)$$

$$B_{\omega x} = -\frac{\omega}{ck_g} E_{\omega y} \,, \qquad (4.125)$$

$$B_{\omega y} = -\frac{\omega}{ck_g} E_{\omega x} \,, \qquad (4.126)$$

and

$$B_{\omega z} = 0 \qquad (4.127)$$

(since this is a TM mode).

Why is this interesting? Let's look again at Equation (4.121). According to this expression for k_g, the $(\text{TM})_{mn}$ mode propagates down the guide or is attenuated, according to whether or not k_g^2 is greater than zero. That is, the exclusionary condition is whether or not the wave frequency ω exceeds the mode frequency ω_{mn}, where

$$\omega_{mn}^2 \equiv c^2 \pi^2 \left(\frac{m^2}{a^2} + \frac{n^2}{b^2} \right) \,. \qquad (4.128)$$

This is also known as the cut-off frequency for the given mode. Thus, for a given frequency ω, the dimensions a and b of the guide can be chosen such that certain modes are excluded. For example, the $(\text{TM})_{11}$ mode will be the only permitted one when $(a^{-2} + b^{-2}) \approx (\omega/c\pi)^2$.

4.6   RADIATION

Thus far in this chapter, we have concentrated on the physics of wave propagation in both bounded (wave guides) and unbounded media. It is now time for us to redirect our attention to the topic of how these electromagnetic waves are generated.

4.6.1   Point Currents and Liénard-Wiechert Potentials

We begin by looking at the potentials and fields produced by point charges, since for them it is possible, at least initially, to define a trajectory $\mathbf{r}_0(t')$ a priori. It must be obvious, however, that when a charge q is radiating, it is giving away momentum and energy, and possibly angular momentum, and that its path cannot therefore be predetermined in the simple fashion we assume here. To adequately address this shortcoming, we must reconsider the physics of radiating systems in Chapter 7, after we have introduced the necessary special relativistic corrections. But for now, we shall proceed with the assumption that when the particle is moving with a velocity much smaller than c, the modifications to its motion resulting from the process of field emission remain inconsequential to the overall behavior of the particle-field system. We shall test the validity of this approach in Chapter 7.

The density of the moving charge is given by

$$\rho(\mathbf{r}', t') = q\,\delta^3(\mathbf{r}' - \mathbf{r}_0[t']) . \tag{4.129}$$

Since in general the current density $\mathbf{J}$ is $\rho\mathbf{v}$, we also have

$$\mathbf{J}(\mathbf{r}', t') = q\mathbf{v}\delta^3(\mathbf{r}' - \mathbf{r}_0[t']) , \tag{4.130}$$

where

$$\mathbf{v}(t') = \frac{d\mathbf{r}_0}{dt'} . \tag{4.131}$$

We recall that in the Lorenz gauge (which is defined by the condition $\vec{\nabla} \cdot \mathbf{A} + [1/c][\partial\Phi/\partial t] = 0$) the potentials satisfy the wave equations (3.13) and (3.14), whose solution is the retarded functions

$$\Phi(\mathbf{r}, t) = \int \frac{\rho(\mathbf{r}', t - |\mathbf{r} - \mathbf{r}'|/c)}{|\mathbf{r} - \mathbf{r}'|}\, d^3x' , \tag{4.132}$$

and

$$\mathbf{A}(\mathbf{r}, t) = \frac{1}{c} \int \frac{\mathbf{J}(\mathbf{r}', t - |\mathbf{r} - \mathbf{r}'|/c)}{|\mathbf{r} - \mathbf{r}'|}\, d^3x' \tag{4.133}$$

(see Equations [3.24] and [3.25] with [3.23]). It is not difficult to see that these *retarded potentials* take into account the finite propagation speed of the electromagnetic disturbance since an effect measured at $\mathbf{r}$ and t was produced at the position of the source at time

$$\tilde{t} = t - \frac{|\mathbf{r} - \mathbf{r}_0(\tilde{t})|}{c} . \tag{4.134}$$

(Quantities labeled with a tilde denote the retarded values.) Thus, using our expressions for ρ and $\mathbf{J}$ (Equations [4.129] and [4.130]) and putting $\vec{\beta} \equiv \mathbf{v}/c$,

$$\Phi(\mathbf{r}, t) = q \int \frac{\delta^3(\mathbf{r}' - \mathbf{r}_0[t - |\mathbf{r} - \mathbf{r}'|/c])}{|\mathbf{r} - \mathbf{r}'|}\, d^3x' , \tag{4.135}$$

and

$$\mathbf{A}(\mathbf{r}, t) = q \int \frac{\vec{\beta}(t - |\mathbf{r} - \mathbf{r}'|/c)\delta^3(\mathbf{r}' - \mathbf{r}_0[t - |\mathbf{r} - \mathbf{r}'|/c])}{|\mathbf{r} - \mathbf{r}'|} \, d^3 x' \ . \qquad (4.136)$$

Note that for a given spacetime field point $(\mathbf{r}, t)$, there exists only *one point* on the whole particle trajectory (Figure 4.7), the retarded coordinate $\tilde{\mathbf{r}}$ corresponding to the retarded time $\tilde{t}$ defined in Equation (4.134) above, which produces a contribution:

$$\tilde{\mathbf{r}} = \mathbf{r}_0 \left(t - \frac{|\mathbf{r} - \tilde{\mathbf{r}}|}{c} \right) \ . \qquad (4.137)$$

In principle, this equation determines $\tilde{\mathbf{r}}$, though in practice it is quite intractable.

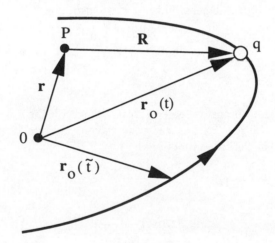

Figure 4.7 The assumed particle trajectory $\mathbf{r}_0(\tilde{t})$ for the charge q, where $\tilde{t}$ is the so-called retarded time, which differs from the observation time t due to the finite light travel time from the emission to the detection points. The observation point is P at $\mathbf{r}$.

To begin evaluating the integrals in Equations (4.135) and (4.136), let us now define the vector

$$\mathbf{R}(t') = \mathbf{r} - \mathbf{r}_0(t') \qquad (4.138)$$

in the direction $\hat{n} \equiv \mathbf{R}/R$. Then,

$$\Phi(\mathbf{r}, t) = q \int \frac{\delta^3(\mathbf{r}' - \mathbf{r}_0[t - R(t')/c])}{R(t')} \, d^3 x' \ , \qquad (4.139)$$

and

$$\mathbf{A}(\mathbf{r}, t) = q \int \frac{\vec{\beta}(t - R(t')/c)\delta^3(\mathbf{r}' - \mathbf{r}_0[t - R(t')/c])}{R(t')} \, d^3x' \, . \tag{4.140}$$

Because the integration variable $\mathbf{x}'$ appears in $R(t')$, we need to transform it, and we do this by introducing the new parameter $\mathbf{r}^*$, where

$$\mathbf{r}^* = \mathbf{r}' - \mathbf{r}_0 \left[t - \frac{R(t')}{c} \right] \, . \tag{4.141}$$

The volume elements d^3x^* and d^3x' are related by the Jacobian transformation

$$d^3x^* = J \, d^3x' \, , \tag{4.142}$$

where

$$J \equiv [1 - \hat{n}(t') \cdot \vec{\beta}(t')] \tag{4.143}$$

is the Jacobian. Heuristically, one can see how the right-hand side of this definition results from "differentiating" $\mathbf{r}^*$ in (4.141) with respect to $\mathbf{r}'$, since the $\mathbf{r}'$ term then $\to 1$ whereas by the chain rule of differentiation, the $\mathbf{r}_0 [t - R(t')/c]$ term $\to -\hat{n}(t') \cdot \vec{\beta}(t')$. With the new integration variable, the integrals for the potential therefore transform to

$$\Phi(\mathbf{r}, t) = q \int \frac{\delta^3(\mathbf{r}^*) \, d^3x^*}{|\mathbf{r} - \mathbf{r}^* - \mathbf{r}_0(\tilde{t})|(1 - \hat{n} \cdot \vec{\beta})} \, , \tag{4.144}$$

and

$$\mathbf{A}(\mathbf{r}, t) = q \int \frac{\vec{\beta}(\tilde{t})\delta^3(\mathbf{r}^*) \, d^3x^*}{|\mathbf{r} - \mathbf{r}^* - \mathbf{r}_0(\tilde{t})|(1 - \hat{n} \cdot \vec{\beta})} \, , \tag{4.145}$$

which can be evaluated trivially, since the argument of the delta function restricts $\mathbf{r}^*$ to a single value:

$$\Phi(\mathbf{r}, t) = \left[\frac{q}{(1 - \hat{n} \cdot \vec{\beta})|\mathbf{r} - \tilde{\mathbf{r}}|} \right]_{\tilde{t}} = \left[\frac{q}{(1 - \hat{n} \cdot \vec{\beta})R} \right]_{\tilde{t}} \, , \tag{4.146}$$

$$\mathbf{A}(\mathbf{r}, t) = \left[\frac{q\vec{\beta}}{(1 - \hat{n} \cdot \vec{\beta})|\mathbf{r} - \tilde{\mathbf{r}}|} \right]_{\tilde{t}} = \left[\frac{q\vec{\beta}}{(1 - \hat{n} \cdot \vec{\beta})R} \right]_{\tilde{t}} \, . \tag{4.147}$$

These are known as the Liénard-Wiechert potentials. Before we go on and use these expressions to evaluate $\mathbf{E}$ and $\mathbf{B}$, let us take a moment to think about the physical meaning of the term $(1 - \hat{n} \cdot \vec{\beta})$, which clearly arises from the fact that the velocity of electromagnetic waves is finite, so that retardation effects must be taken into account in determining the fields.

Special Note. We can understand the meaning of the "shrinkage" factor $(1 - \hat{n} \cdot \vec{\beta})$ by considering a thin cylinder of charge moving along the x-axis with velocity v (Figure 4.8). To calculate the field at x when the ends of the cylinder are at (x_1, x_2), we need to know the location of the retarded points $\tilde{x}_1$ and $\tilde{x}_2$:

$$x_1 - \tilde{x}_1 = \frac{v}{c}(x - \tilde{x}_1) \,, \tag{4.148}$$

$$x_2 - \tilde{x}_2 = \frac{v}{c}(x - \tilde{x}_2) \,. \tag{4.149}$$

Thus, with $\tilde{L} \equiv \tilde{x}_2 - \tilde{x}_1$, we can subtract Equation (4.149) from (4.148) to get

$$\tilde{L} - L = \frac{v}{c}\tilde{L} \,, \tag{4.150}$$

or

$$\tilde{L} = \frac{L}{1 - v/c} \,. \tag{4.151}$$

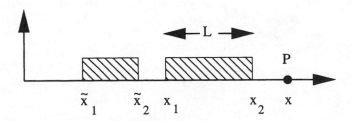

Figure 4.8 A one-dimensional (heuristic) problem to demonstrate the role of the "shrinkage" factor when the emitter is traveling relative to the observer.

That is, the effective length $\tilde{L}$ differs by the factor $(1 - \hat{x} \cdot \vec{\beta})^{-1} = (1 - v/c)^{-1}$ compared to the natural length L because the source is moving relative to the observer and its velocity must be taken into account when calculating the retardation effects.

4.6.2   The Radiation Fields

Now that we have the potentials (Equations [4.146] and [4.147]), calculating the fields should be straightforward with the use of the defining Equations (1.35) and (1.37). In fact, however, it turns out to be easier to first take a step back and instead write the Liénard-Wiechert potentials in the equivalent form

$$\Phi(\mathbf{r}, t) = q \int \frac{\delta[t' - t + R(t')/c]}{R(t')} \, dt' \, , \tag{4.152}$$

and

$$\mathbf{A}(\mathbf{r}, t) = q \int \frac{\vec{\beta}(t')\delta[t' - t + R(t')/c]}{R(t')} \, dt' \, , \tag{4.153}$$

where $R(t') \equiv |\mathbf{r} - \mathbf{r}_0(t')|$. This can be verified easily using the following property of the Dirac delta function (e.g., Davydov 1976):

$$\int g(x)\,\delta[f(x)]\,dx = \sum_i \left[\frac{g(x)}{|df/dx|}\right]_{f(x_i)=0} \, , \tag{4.154}$$

which holds for regular functions $g(x)$ and $f(x)$ of the integration variable x, where x_i are the zeros of $f(x)$. The advantage in pursuing this path is that the derivatives in Equations (1.35) and (1.37) can be carried out before the integration over the delta function, which simplifies the evaluation of the fields considerably since, for one thing, we do not need to keep track of the retarded time until the last step. We get for the electric field

$$\mathbf{E}(\mathbf{r}, t) = -q \int \vec{\nabla} \left[\frac{\delta(t' - t + R(t')/c)}{R(t')}\right] \, dt'$$

$$-\frac{q}{c}\frac{\partial}{\partial t} \int \frac{\vec{\beta}(t')\,\delta(t' - t + R(t')/c)}{R(t')} \, dt' \, . \tag{4.155}$$

Thus, differentiating the integrand in the first term, we get

$$\mathbf{E}(\mathbf{r}, t) = q \int \left[\frac{\hat{n}}{R^2}\,\delta\left(t' - t + \frac{R(t')}{c}\right) - \frac{\hat{n}}{cR}\,\delta'\left(t' - t + \frac{R(t')}{c}\right)\right] \, dt'$$

$$-\frac{q}{c}\frac{\partial}{\partial t} \int \frac{\vec{\beta}(t')\,\delta(t' - t + R(t')/c)}{R(t')} \, dt' \, . \tag{4.156}$$

But

$$\delta'\left(t' - t + \frac{R(t')}{c}\right) = -\frac{\partial}{\partial t}\,\delta\left(t' - t + \frac{R(t')}{c}\right) \, , \tag{4.157}$$

and so

$$\mathbf{E}(\mathbf{r}, t) = q \int \frac{\hat{n}}{R^2}\,\delta\left(t' - t + \frac{R(t')}{c}\right) \, dt'$$

$$+q\frac{\partial}{\partial t} \int \frac{(\hat{n} - \vec{\beta})}{cR(t')}\,\delta\left(t' - t + \frac{R(t')}{c}\right) \, dt' \, . \tag{4.158}$$

It's clear that the next step is to evaluate the integrals using the property of the Dirac delta function expressed in Equation (4.154), but to do that, we need to know the derivative of the delta function's argument with respect to t'. By the chain rule of differentiation,

$$\frac{d}{dt'}\left(t' - t + \frac{R(t')}{c}\right) = (1 - \hat{n} \cdot \vec{\beta})_{\tilde{t}} ,\qquad(4.159)$$

with which we then easily get the result

$$\mathbf{E}(\mathbf{r}, t) = q\left\{\frac{\hat{n}}{(1 - \hat{n} \cdot \vec{\beta})R^2}\right\}_{\tilde{t}} + \frac{q}{c}\frac{\partial}{\partial t}\left\{\frac{\hat{n} - \vec{\beta}}{(1 - \hat{n} \cdot \vec{\beta})R}\right\}_{\tilde{t}} .\qquad(4.160)$$

In the last step of this rather long derivation, we evaluate the derivative on the right-hand side of this expression using the known relationship between t and $\tilde{t}$ (Equation [4.134]). Since $\partial R/\partial t = (\partial R/\partial t')(\partial t'/\partial t) = -\hat{n} \cdot \mathbf{v}(\partial t'/\partial t) = c(1 - \partial t'/\partial t)$, it is not difficult to see that

$$\frac{\partial \tilde{t}}{\partial t} = \frac{1}{(1 - \hat{n} \cdot \vec{\beta})} .\qquad(4.161)$$

Thus,

$$\frac{\partial}{\partial t}\left\{\frac{\hat{n} - \vec{\beta}}{(1 - \hat{n} \cdot \vec{\beta})R}\right\}_{\tilde{t}} = \frac{1}{(1 - \hat{n} \cdot \vec{\beta})}\frac{\partial}{\partial \tilde{t}}\left\{\frac{\hat{n} - \vec{\beta}}{(1 - \hat{n} \cdot \vec{\beta})R}\right\}_{\tilde{t}} ,\qquad(4.162)$$

and using the additional pieces

$$\dot{R}|_{\tilde{t}} = -c\,(\hat{n} \cdot \vec{\beta})_{\tilde{t}} ,\qquad(4.163)$$

$$\dot{\hat{n}}|_{\tilde{t}} = \frac{c}{R}\,[\hat{n}(\hat{n} \cdot \vec{\beta}) - \vec{\beta}]_{\tilde{t}} ,\qquad(4.164)$$

and

$$\frac{d}{d\tilde{t}}(1 - \hat{n} \cdot \vec{\beta})|_{\tilde{t}} = -(\hat{n} \cdot \dot{\vec{\beta}} + \vec{\beta} \cdot \dot{\hat{n}}) ,\qquad(4.165)$$

we get finally,

$$\mathbf{E}(\mathbf{r}, t) = q\left\{\frac{(\hat{n} - \vec{\beta})(1 - \beta^2)}{(1 - \hat{n} \cdot \vec{\beta})^3 R^2} + \frac{\hat{n} \times [(\hat{n} - \vec{\beta}) \times \dot{\vec{\beta}}]}{c(1 - \hat{n} \cdot \vec{\beta})^3 R}\right\}_{\tilde{t}} .\qquad(4.166)$$

A similar procedure for $\mathbf{B}$ shows that

$$\mathbf{B}(\mathbf{r}, t) = \vec{\nabla} \times \mathbf{A} = \hat{n}(\tilde{t}) \times \mathbf{E} .\qquad(4.167)$$

In analyzing this elegant equation, we immediately gain several physical insights. First, we confirm that when the particle is at rest and unaccelerated with respect to us, the field reduces simply to the Coulomb's law $q\hat{n}/R^2$ that we have come to expect. Whatever corrections are introduced to the field from the motion and/or acceleration of the source, they do not alter the limiting empirical law that we know is correct. As $\beta \to 1$ with $\dot{\beta} = 0$, however, the field displays a "bunching" effect, not unlike what happens to sound waves when the flow velocity in a gas approaches the sonic value and creates a shock front. Looking at the first term in Equation (4.166), it is clear that in this case $\mathbf{E} \to 0$ for all directions except $\hat{n} = \hat{\beta}$. In this limit, of course, the source is moving with a velocity approaching that of the field it is producing, and it should not surprise us that the field lines at the observation point should then have a higher density than is the case for a static radial field. It is important to emphasize that in deriving this expression for $\mathbf{E}$, we did not have any knowledge of special relativity (which is discussed in the next chapter). Yet, as we shall see in Chapter 7, Equation (4.166) turns out to be correct even for $\beta \to 1$, but we will then better understand why β cannot exceed 1. As such, this "bunching" is understood as being a retardation effect, resulting from the finite velocity of electromagnetic waves. For now, we will work under the premise that Equation (4.166) is valid at least when $\beta \ll 1$ (i.e., the low velocity limit).

We also see a clean separation into the *near* field (which falls off as R^{-2}) and the far or *radiation* field (which falls off as R^{-1}). Unless the particle is accelerated, so that $\dot{\beta} \neq 0$, the field falls off rapidly at large distances. But when a radiation field is present, it dominates over the near field far from the source. We shall quantify these concepts of relative distance and the character of the field in the next subsection, and we shall return to this topic in Chapter 7 where we demonstrate that only the radiation field carries energy and momentum away from the charge, constituting a dynamical transport that points to a separation of the field from its source.

Example 4.5. When the motion of the source is uniform, so that $\ddot{\beta} = 0$, the electric field contains only the near field term:

$$\mathbf{E}_{\text{near}}(\mathbf{r}, t) = q \left\{ \frac{(\hat{n} - \vec{\beta})(1 - \beta^2)}{(1 - \hat{n} \cdot \vec{\beta})^3 R^2} \right\}_{\tilde{t}} . \tag{4.168}$$

Let this motion be along the x-axis. We wish to calculate the field at P (in Figure 4.9) when the charge is at Q. In this geometry,

$$\hat{n}(\tilde{t}) - \vec{\beta} = \frac{\mathbf{R}(\tilde{t}) - \vec{\beta}\,R(\tilde{t})}{R(\tilde{t})} = \frac{\mathbf{R}(t)}{R(\tilde{t})} \ . \tag{4.169}$$

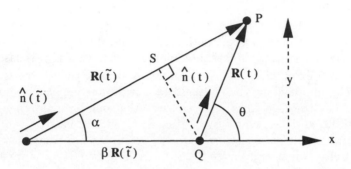

Figure 4.9 The geometry of uniform particle motion with velocity $\mathbf{v} = \vec{\beta}c$ in the direction $\hat{x}$, relative to an observer at P. The y-axis points straight up and it is to be noted that θ measures the angle of $\mathbf{R}(t)$ relative to $\hat{x}$ at the time the measurement is made, not the time $(= \tilde{t})$ when the particle produced the field.

In addition,

$$(1 - \hat{n} \cdot \vec{\beta})_{\tilde{t}} = (1 - \beta \cos \alpha) \ , \tag{4.170}$$

and since $\bar{S}P = R(\tilde{t})(1 - \beta \cos \alpha)$, we see that

$$R(\tilde{t})(1 - \beta \cos \alpha) = \left\{ R^2(t) - \beta^2 R^2(t) \sin^2 \theta \right\}^{1/2} \ . \tag{4.171}$$

Thus, all told,

$$\mathbf{E}_{\text{near}}(\mathbf{r}, t) = q \left\{ \frac{\mathbf{R}(t)(1 - \beta^2)}{R^3(t)(1 - \beta^2 \sin^2 \theta)^{3/2}} \right\} \ . \tag{4.172}$$

We confirm again that when $\beta = 0$ we recover Coulomb's law, since

$$\mathbf{E}_{\text{near}}(\mathbf{r}, t) \to \mathbf{E}_C(\mathbf{r}, t) \equiv \frac{q\mathbf{R}(t)}{R^3(t)} \ . \tag{4.173}$$

But notice that $|\mathbf{E}_{\text{near}}| < |\mathbf{E}_C|$ below some critical angle θ_C depending on the value of β. For example, $\theta_C \approx 65°$ when $\beta = 0.9$. The reason for this is that the particle was actually further than it appears to be from a Galilean perspective when it produced the field observed at the current time. On the other hand, $|\mathbf{E}_{\text{near}}| > |\mathbf{E}_C|$ for $\theta > \theta_C$ due to the "bunching" of field lines discussed above.

Example 4.6. Since the velocity term drops off with distance as R^{-2}, whereas the radiation term goes as R^{-1}, the former is of little practical interest far from the source, i.e., in the "radiation zone." It is common, therefore, to talk of a radiation field, which is simply the second term in Equation (4.166):

$$\mathbf{E}_{\text{rad}}(\mathbf{r}, t) = \frac{q}{c} \left\{ \frac{\hat{n} \times [(\hat{n} - \vec{\beta}) \times \dot{\vec{\beta}}]}{(1 - \hat{n} \cdot \vec{\beta})^3 R} \right\}_{\tilde{t}} . \tag{4.174}$$

Let us now see what happens when the acceleration of the charge is linear, in which case $\vec{\beta} \times \dot{\vec{\beta}} = 0$, and let us assume for simplicity that $\dot{\vec{\beta}}$ is constant ($= a/c$) during a small segment of the particle's motion, and zero otherwise. During this segment, lasting a time Δt, the particle emits a pulse of radiation that will be felt at some distant point over a corresponding, limited time interval.

We have stressed that our derivation of Equation (4.166) assumes nonrelativistic motion, and so (4.174) is valid only in the low velocity limit. If in fact $\beta \ll 1$, then $(1 - \hat{n} \cdot \vec{\beta}) \to 1$, and letting θ be the angle between $\hat{n}$ and $\vec{\beta}$, we easily reduce (4.174) to the simpler form

$$\mathbf{E}_{\text{rad}}(\mathbf{r}, t) = \frac{qa \sin \theta}{c^2 r} \hat{\theta} . \tag{4.175}$$

The Poynting vector (Equation [3.44]) for this field has a magnitude

$$S = \frac{c}{4\pi} |\mathbf{E}_{\text{rad}}|^2 = \frac{q^2 a^2 \sin^2 \theta}{4\pi c^3 r^2} , \tag{4.176}$$

which therefore represents the characteristic radiation pattern ($\propto \sin^2 \theta$) in the dipole approximation (Figure 4.10).

4.6.3 Simple Radiating Systems

When we move away from individual radiating charges, which have so far commanded our attention for most of this chapter, calculating the field structure can become a challenging task, particularly for complicated density (ρ) and current ($\mathbf{J}$) distributions. In cases such as this, it is essential that we introduce a simplifying procedure, the most common of which is the expansion of the relevant quantities as Fourier series. That is, for a given scalar or vector component function $X(\mathbf{x}, t)$, we write

$$X(\mathbf{x}, t) = \sum_{\omega} X_{\omega}(\mathbf{x}) \exp(-i\omega t) . \tag{4.177}$$

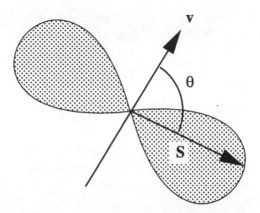

Figure 4.10 The characteristic radiation pattern in the dipole approximation. The lobes indicate the loci of the tips of the vector **S** as a function of
angle θ relative to the particle's direction of motion. This is a cross-sectional
cut, so the complete surface is toroidal around **v**.

The hope is that one can then solve for each of the frequency components
separately; indeed, the solution may be dominated by only a handful of these
terms. Let us begin with the retarded potential solutions in the Lorenz gauge
(Equations [4.132] and [4.133]), which are conveniently written in terms of
general ρ and **J** profiles. The Lorenz gauge forces the scalar potential to satisfy
the condition

$$\vec{\nabla} \cdot \mathbf{A}_\omega - ik\Phi_\omega = 0 , \qquad (4.178)$$

where $k = \omega/c$, and

$$\mathbf{A}_\omega = \frac{1}{c} \int \mathbf{J}_\omega(\mathbf{x}') \frac{\exp(ik|\mathbf{x} - \mathbf{x}'|)}{|\mathbf{x} - \mathbf{x}'|} \, d^3x' , \qquad (4.179)$$

and so both the electric and magnetic fields are derivable from the vector field
alone (see Equations [1.35] and [1.37]):

$$\mathbf{E}_\omega = \frac{i}{k}\vec{\nabla}(\vec{\nabla} \cdot \mathbf{A}_\omega) + ik\mathbf{A}_\omega , \qquad (4.180)$$

$$\mathbf{B}_\omega = \vec{\nabla} \times \mathbf{A}_\omega . \qquad (4.181)$$

Look carefully at Equation (4.179), because this is the key to the whole solution.
There are evidently three length scales involved here: (i) the wavelength $\lambda = 2\pi/k$, (ii) the distance $r = |\mathbf{x}|$ to the field point, and (iii) the size (let's call it

D) of the source. In principle, this suggests that the nature of the solution can be characterized in at least three different regimes. Here we consider the two most interesting and relevant ones, which we shall call the "near" and "far" zones, to be quantified below. The solution separates cleanly into the different limiting forms because of the expandability of the term

$$\frac{\exp(ik|\mathbf{x} - \mathbf{x}'|)}{|\mathbf{x} - \mathbf{x}'|} . \tag{4.182}$$

For $r > r'$, with γ the angle between $\mathbf{x}$ and $\mathbf{x}'$, we have

$$k|\mathbf{x} - \mathbf{x}'| = kr\sqrt{1 + \frac{r'^2}{r^2} - 2\frac{r'}{r}\cos\gamma} , \tag{4.183}$$

so that

$$k|\mathbf{x} - \mathbf{x}'| \approx kr\left\{1 + \frac{1}{2}\left(\frac{r'^2}{r^2} - \frac{2r'}{r}\cos\gamma\right) - \frac{1}{8}\left(\frac{r'^2}{r^2} - \frac{2r'}{r}\cos\gamma\right)^2 \ldots\right\} , \tag{4.184}$$

and therefore

$$\exp(ik|\mathbf{x} - \mathbf{x}'|) \approx \exp\{ikr + (ikr'/2)(r'/r) - ikr'\cos\gamma\} . \tag{4.185}$$

In the *near* (or quasi-stationary) zone, defined by the condition $kr \ll 1$, the answer is quite straightforward since

$$\exp(ik|\mathbf{x} - \mathbf{x}'|) \approx 1 , \tag{4.186}$$

and in this case

$$\mathbf{A}_\omega \approx \frac{1}{c}\int \frac{\mathbf{J}_\omega(\mathbf{x}')\,d^3x'}{|\mathbf{x} - \mathbf{x}'|} . \tag{4.187}$$

Formally, this looks just like the expression for $\mathbf{A}$ in magnetostatics, where the current is stationary. Here, however, $\mathbf{J}$ is introduced as a time-dependent variable, though we consider the (near) regime where time retardation is negligible. In this zone, $\lambda = 2\pi/k$ is much larger than r, which means that $\mathbf{E}$ and $\mathbf{B}$ are calculable as in the static case because $\mathbf{B}$ varies with the *same* phase as the currents and $\mathbf{E}$ with the same phase as the charge—all typical radiation effects are absent because the fields have not yet *decoupled* from the sources.

In the *far* (or radiation) zone, defined by $r \gg D$ and $r \gg \lambda$, we have

$$\frac{\exp(ik|\mathbf{x} - \mathbf{x}'|)}{|\mathbf{x} - \mathbf{x}'|} \approx \frac{\exp(ikr)}{r}\exp(-ikr'\cos\gamma) , \tag{4.188}$$

and so

$$\mathbf{A}_\omega \approx \frac{\exp(ikr)}{cr} \int \mathbf{J}_\omega \exp(-ikr' \cos\gamma) \, d^3x' \, . \tag{4.189}$$

Now, in solving for the fields, we will need

$$\vec{\nabla} \cdot \mathbf{A}_\omega = \left[-\frac{\mathbf{r}}{r} \frac{\exp(ikr)}{r^2} + \frac{ik\exp(ikr)}{r} \frac{\mathbf{r}}{r} \right]$$

$$\times \frac{1}{c} \int \mathbf{J}_\omega \exp(-ikr' \cos\gamma) \, d^3x' \, . \tag{4.190}$$

But since $kr \gg 1$, we obviously have

$$\vec{\nabla} \cdot \mathbf{A}_\omega = ik(\hat{r} \cdot \mathbf{A}_\omega) \, , \tag{4.191}$$

and making the same approximation with $\vec{\nabla}(\vec{\nabla} \cdot \mathbf{A}_\omega)$, we find that

$$\vec{\nabla}(\vec{\nabla} \cdot \mathbf{A}_\omega) = ik\hat{r}(\vec{\nabla} \cdot \mathbf{A}_\omega) = (ik)^2 \hat{r}(\hat{r} \cdot \mathbf{A}_\omega) \, . \tag{4.192}$$

Thus,

$$\mathbf{E}_\omega = -ik[\hat{r}(\hat{r} \cdot \mathbf{A}_\omega) - \mathbf{A}_\omega] \, . \tag{4.193}$$

This expression for the electric field is rather intriguing, since it suggests that we should subtract from $\mathbf{A}_\omega$ its projection along the line of sight. To pursue this further, let's write

$$\mathbf{A}_\omega = \mathbf{A}_\omega^\perp + \mathbf{A}_\omega^\| \tag{4.194}$$

for the components of $\mathbf{A}_\omega$ perpendicular to and parallel to the direction of observation $\hat{r}$. Here,

$$\mathbf{A}_\omega^\| = \hat{r}(\hat{r} \cdot \mathbf{A}_\omega) \, , \tag{4.195}$$

so that

$$\mathbf{A}_\omega^\perp = -\{\hat{r}(\hat{r} \cdot \mathbf{A}_\omega) - \mathbf{A}_\omega\} \, . \tag{4.196}$$

As such,

$$\mathbf{E}_\omega = ik\mathbf{A}_\omega^\perp \, , \tag{4.197}$$

whence

$$\mathbf{E}_\omega = ik\frac{\exp(ikr)}{cr} \int \mathbf{J}_\omega^\perp(\mathbf{x}') \exp(-ikr' \cos\gamma) \, d^3x' \, . \tag{4.198}$$

We conclude that the electric field is generated from the perpendicular component of the current density only, a result that will prove to be quite useful in practice when this equation is solved for the field in real situations. Finally, since $\hat{r} \times \hat{r} = 0$,

$$\mathbf{B}_\omega = \vec{\nabla} \times \mathbf{A}_\omega = ik\frac{\mathbf{r}}{r} \times \mathbf{A}_\omega = (\hat{r} \times \mathbf{E}_\omega) \, . \tag{4.199}$$

The Poynting vector is here calculated from

$$\mathbf{S} = \frac{c}{8\pi} \operatorname{Re}\left\{\mathbf{E}_\omega \times \mathbf{B}_\omega^*\right\} , \qquad (4.200)$$

which in the end reduces conveniently to a function of the current density only:

$$\mathbf{S} = \frac{k^2}{8\pi c r^2} \left\{ \int \mathbf{J}_\omega^\perp(\mathbf{x}') \exp(-ikr' \cos\gamma)\, d^3 x' \right\}^2 \hat{r} . \qquad (4.201)$$

Its use is primarily to convey the angular distribution of the radiation pattern when the source $\mathbf{J}$ is given.

Example 4.7. One of the simplest applications of the equations we have just derived is the calculation of the radiation produced by a current-carrying thin conductor aligned with, say, the z-axis (Figure 4.11).

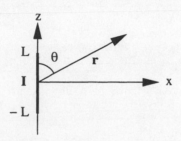

Figure 4.11 Linear antenna, i.e., a thin conductor carrying a current I in the $\hat{z}$ direction.

The electric field produced by this *linear antenna* is given by Equation (4.198) with

$$\mathbf{J}_\omega^\perp = -\hat{r}(\hat{r} \cdot \mathbf{J}_\omega) + \mathbf{J}_\omega , \qquad (4.202)$$

and

$$\cos\gamma = \frac{\mathbf{r} \cdot \mathbf{r}'}{rr'} . \qquad (4.203)$$

In this example,

$$d^3 x'\, \mathbf{J}_\omega^\perp(\mathbf{x}') = \mathbf{I}_\omega^\perp(\mathbf{x}')\, dz' , \qquad (4.204)$$

where $\mathbf{I}_\omega$ is the vector current running along the conductor. That is,

$$d^3 x'\, \mathbf{J}_\omega^\perp(\mathbf{x}') = \left[-\hat{r}\cos\theta\, I_\omega(\mathbf{x}') + \hat{z} I_\omega(\mathbf{x}')\right] dz' = -\hat{\theta} I_\omega \sin\theta\, dz' . \qquad (4.205)$$

Also, $\cos \gamma = \cos \theta$, and therefore

$$\mathbf{E}_\omega = -\frac{ik \exp(ikr)}{cr} \sin \theta \int I_\omega(z') \exp(-ikz' \cos \theta) \, dz' \, \hat{\theta} \, . \tag{4.206}$$

In addition, since $\mathbf{B}_\omega = \hat{r} \times \mathbf{E}_\omega$, it is clear that

$$\mathbf{B}_\omega = |\mathbf{E}_\omega| \, \hat{\phi} \, . \tag{4.207}$$

For simplicity, let us suppose that $I_\omega(z') = I_0$ for $-L \le z' \le L$, and zero otherwise. Then,

$$\mathbf{E}_\omega = -\frac{ik \exp(ikr)}{cr} \sin \theta \, I_0 \left(\frac{2}{k \cos \theta} \right) \sin(kL \cos \theta) \, , \tag{4.208}$$

and from Equation (4.201),

$$|\mathbf{S}| = \frac{k^2}{8\pi cr^2} \sin^2 \theta \, I_0^2 \, \frac{4 \sin^2(kL \cos \theta)}{k^2 \cos^2 \theta} = \frac{I_0^2 \tan^2 \theta \, \sin^2(kL \cos \theta)}{2\pi cr^2} \, . \tag{4.209}$$

Thus, if $L = m\lambda$,

$$|\mathbf{S}| = \frac{I_0^2 \tan^2 \theta \, \sin^2(2\pi m \cos \theta)}{2\pi cr^2} \, , \tag{4.210}$$

and the radiation patterns in Figure 4.12 show the strong dependence on m.

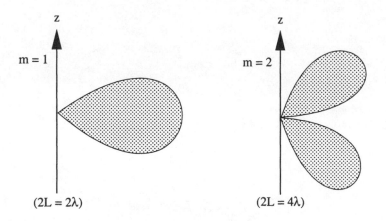

Figure 4.12 Radiation patterns for different modes m from a linear antenna of size $L = m\lambda$, where λ is the wavelength of the radiation.

Example 4.8. An interesting problem is the radiation from an electric dipole, which in the simplest manifestation is simply two opposite charges q and $-q$ separated by a distance l, generating a dipole moment defined to be

$$\mathbf{p} = q\mathbf{l} \, . \tag{4.211}$$

When the charge configuration is instead $\rho(\mathbf{x})$, we have

$$\mathbf{p} \equiv \int \mathbf{x}' \, \rho(\mathbf{x}') \, d^3 x' \, . \tag{4.212}$$

Suppose we look at the long wavelength limit, where $\lambda \gg D$ (the source size), so that $kr' \ll 1$. Then

$$\mathbf{A}_\omega \approx \frac{\exp(ikr)}{cr} \int \mathbf{J}_\omega \, d^3 x' \, . \tag{4.213}$$

But

$$\int \mathbf{J}_\omega \, d^3 x' = -\int \mathbf{x}'(\vec{\nabla}' \cdot \mathbf{J}_\omega) \, d^3 x' \tag{4.214}$$

after integrating by parts, and so

$$\int \mathbf{J}_\omega \, d^3 x' = -i\omega \int \mathbf{x}' \, \rho(\mathbf{x}') \, d^3 x' \, , \tag{4.215}$$

because from the continuity of charge equation

$$\vec{\nabla} \cdot \mathbf{J} + \frac{\partial \rho}{\partial t} = 0 \, , \tag{4.216}$$

which gives

$$\vec{\nabla} \cdot \mathbf{J} = i\omega\rho \, . \tag{4.217}$$

Thus,

$$\mathbf{A}_\omega \approx -\frac{\exp(ikr)}{r} \, ik \, \mathbf{p} \, . \tag{4.218}$$

The radiation field for $\lambda \gg D$ is thus expected to be

$$\mathbf{E}_\omega = -ik[\hat{r}(\hat{r} \cdot \mathbf{A}_\omega) - \mathbf{A}_\omega] = k^2 \frac{\exp(ikr)}{r} \, [\hat{r}(\hat{r} \cdot \mathbf{p}) - \mathbf{p}] \, , \tag{4.219}$$

which shows the typical behavior of a radiation field oscillating as $\exp(ikr)$ and falling off with distance as r^{-1}.

5

A NEED FOR THE SPECIAL THEORY OF RELATIVITY

5.1   BASIC PRINCIPLES AND TRANSFORMATIONS

In spite of its many strengths, the theory of electrodynamics that we have been developing thus far does not satisfy the principles of Galilean relativity. There are several fundamental reasons why this is unacceptable, not the least of which is its apparent inconsistency with the rest of classical mechanics.

A Galilean transformation connects two frames of reference moving relative to each other with a *constant* velocity $\mathbf{v}_o$ (Figure 5.1). The coordinates and time in these two frames are related by

$$\mathbf{x}' = \mathbf{x} - \mathbf{v}_o\, t \tag{5.1}$$

and

$$t' = t \; . \tag{5.2}$$

As is well known, all the physical laws of *classical* mechanics are invariant under such Galilean transformations. For example, Newton's second law of motion in the moving frame is

$$
\begin{aligned}
\mathbf{F}(\mathbf{x}') &= m\,\mathbf{a}' \\
&= m\,\frac{d^2\,\mathbf{x}'}{dt'^2} \\
&= m\,\mathbf{a} - m\,\frac{d\,\mathbf{v}_o}{dt'} \\
&= m\,\mathbf{a} \\
&= \mathbf{F}(\mathbf{x}) \; ,
\end{aligned}
\tag{5.3}
$$

where m and **a** are the particle's mass and acceleration, respectively, showing explicitly that its mathematical *form* in terms of the relevant physical parameters is preserved, regardless of which (unaccelerated) reference frame is being considered.

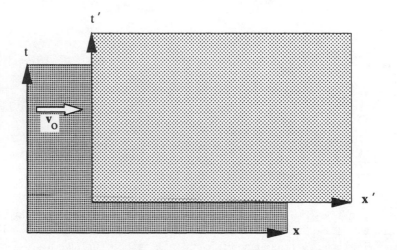

Figure 5.1 Two Cartesian coordinate frames moving relative to each other with a constant velocity v_0. By convention, the unprimed system is taken to be fixed to the laboratory. In this figure, the three-dimensional space is represented by the horizontal axis, whereas time is measured along the vertical axis.

It is essential that electrodynamics also be structured with physical laws that are invariant under transformations between reference frames moving with a constant velocity relative to each other. This need was first brought into sharp focus by the results of the Michelson-Morley experiment (Michelson and Morley 1887), which showed that the velocity of light is the same, within 5 km/sec, for light traveling along the direction of the earth's orbital motion and transverse to it, even though the earth has a velocity of about 30 km/sec relative to the sun, and about 200 km/s relative to the center of our galaxy. The accuracy of this result has more recently been improved to about 1 km/s by Jaseva, Javan, Murray, and Townes (1964). In its present form, our theory of electrodynamics predicts that the speed of light in vacuum is a universal constant, c, but if this is true in one coordinate system $(\mathbf{x}, t)$ then it is not true in the "moving" coordinate system $(\mathbf{x}', t')$ defined by the Galilean transformation (Equations [5.1] and [5.2]). Either the theory is wrong—which is unlikely given that the four principal equations are merely statements of experimental fact—or the Galilean relativity principle is incomplete.

The fact that our current theory of electrodynamics is not invariant was already evident in our discussion of Faraday's law through its application to a moving loop C (Figure 5.2):

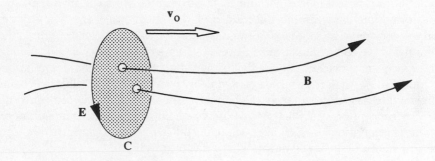

Figure 5.2 The application of Faraday's law to a loop C moving with constant velocity $\mathbf{v}_0$ through a background magnetic field $\mathbf{B}$.

In the loop's comoving frame, we have

$$\oint_C \mathbf{E}' \cdot d\mathbf{l} = -\frac{1}{c}\frac{d\Phi'_B}{dt'}$$

$$= -\frac{1}{c}\oint_S \frac{d\mathbf{B}'}{dt'} \cdot d\mathbf{a}$$

$$= -\frac{1}{c}\oint_S \frac{\partial\mathbf{B}'}{\partial t'} \cdot d\mathbf{a}, \qquad (5.4)$$

where Φ'_B is the magnetic flux threading the loop, and S is the area bounded by C. (Note that $d/dt' = \partial/\partial t' + \mathbf{v}\cdot\vec{\nabla}' = \partial/\partial t'$ here, since $\mathbf{v} = 0$ in this frame.) But in the laboratory frame, Faraday's law becomes

$$\oint_C \mathbf{E} \cdot d\mathbf{l} = -\frac{1}{c}\oint_S \left(\frac{\partial\mathbf{B}}{\partial t} + \mathbf{v}_o\cdot\vec{\nabla}\mathbf{B}\right) \cdot d\mathbf{a}, \qquad (5.5)$$

where now

$$(\mathbf{v}_o \cdot \vec{\nabla})\mathbf{B} = \vec{\nabla}\times(\mathbf{B}\times\mathbf{v}_o) + \mathbf{v}_o(\vec{\nabla}\cdot\mathbf{B}) - \mathbf{B}(\vec{\nabla}\cdot\mathbf{v}_o) + (\mathbf{B}\cdot\vec{\nabla})\mathbf{v}_o$$

$$= \vec{\nabla}\times(\mathbf{B}\times\mathbf{v}_o). \qquad (5.6)$$

That is,

$$\oint_C \left(\mathbf{E} - \frac{1}{c} \mathbf{v}_o \times \mathbf{B} \right) \cdot d\mathbf{l} = -\frac{1}{c} \oint_S \frac{\partial \mathbf{B}}{\partial t} \cdot d\mathbf{a} . \tag{5.7}$$

Thus, there is a net change in the *form* of the electric field **E**, *that depends on the relative velocity* $\mathbf{v}_o$, due to the transformation from one frame to the other. This physical law—one of the cornerstones of electrodynamics—is not invariant under a Galilean transformation! We are forced to concede the fact that if both classical mechanics and electrodynamics are to be invariant under transformations between inertial frames of reference, either the Galilean relativity theory must be modified, or we need an alternative understanding of how these fields are to be transformed from one frame to the next. As we shall see, the experimental evidence favors the former, and classical mechanics must itself be modified so that it is no longer Galilean invariant, while electrodynamics remains unchanged in so far as the equations in any one frame are concerned. But our concept of what the electromagnetic field is must be reshaped in order to cast the equations in such a way that both theories are then invariant under the transformations of the new (i.e., special) relativity.

The two postulates upon which we must base the new framework are:

(1) Only *relative* motion is observable, and

(2) The velocity of light in vacuum is a *constant*, c, regardless of the source and/or observer speed.[15]

Because of postulate (2), it is clear that observers in different reference frames will no longer agree on the space and time (*spacetime*) coordinates of an event. For suppose a pulse of light is emitted at time $t = 0$ when the point O is the common origin of the two reference frames (Figure 5.3). The observer at O will report that the pulse arrival time at A is

$$t = r/c , \tag{5.8}$$

where

$$r = (x^2 + y^2 + z^2)^{1/2} . \tag{5.9}$$

The observer at O', on the other hand, will report a corresponding time

$$t' = r'/c , \tag{5.10}$$

where

$$r' = (x'^2 + y'^2 + z'^2)^{1/2} . \tag{5.11}$$

---

[15]For early critical reviews of the experimental basis for the second postulate, see Fox (1965, 1967).

Because the speed of light, c, is the same in both frames, and because the two origins have separated during the propagation of the pulse, it is clear that $r \neq r'$ and $t \neq t'$.

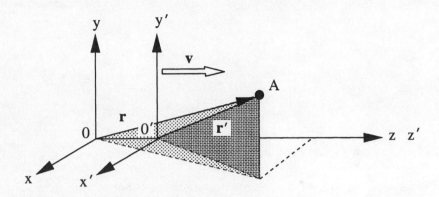

Figure 5.3 The emission of a pulse of light at the origin of the coordinate system as seen by two observers moving with a velocity $\mathbf{v} \parallel \hat{z}$ relative to each other.

To establish a connection between the two coordinate systems, it is often useful to deal with the "spacetime interval"

$$(\Delta s)^2 \equiv (c\,\Delta t)^2 - |\Delta\,\mathbf{x}|^2 \, , \tag{5.12}$$

which is *always* zero along a light path. Thus, in the example we have just described,

$$(\Delta s)^2 = (c\,t)^2 - r^2 = 0 = (c\,t')^2 - r'^2 = (\Delta s')^2 \, . \tag{5.13}$$

Our problem is to find a transformation that preserves the invariance exhibited in (5.13), a process that involves nothing more than guessing the correct relation between the primed and unprimed coordinates. It is not difficult to show that in order for the spacetime interval to be preserved, the coordinates in the two frames must be related via the so-called Lorentz transformation, viz.,

$$
\begin{aligned}
x' &= x \, , \\
y' &= y \, , \\
z' &= \gamma\,(z - v\,t) \, , \\
t' &= \gamma\left(t - \frac{v\,z}{c^2}\right) , \\
\gamma &\equiv \left[1 - \left(\frac{v}{c}\right)^2\right]^{-1/2} .
\end{aligned}
\tag{5.14}
$$

Other transformations preserving the speed of light—for instance, a rigid rotation—are available, and together these constitute the homogeneous Lorentz group. However, only the relations in Equation (5.14) account for the transformation of the coordinates from one frame to another moving relative to it with a constant, uniform velocity. As we should expect, the reciprocal transformation is just

$$
\begin{aligned}
x &= x' \,, \\
y &= y' \,, \\
z &= \gamma \left(z' + v\, t' \right) , \\
t &= \gamma \left(t' + \frac{v\, z'}{c^2} \right) ,
\end{aligned}
\tag{5.15}
$$

since the observer at O' attributes a velocity $-v$ to the frame moving with O.

Suppose now we wanted to measure the distance between two *simultaneous* events, as seen in each frame. Let L_0 be a length measured in the frame moving with respect to us. Then,

$$
L_0 \equiv z_2' - z_1' = \gamma \left(z_2 - z_1 - v\, t_2 + v\, t_1 \right) ,
\tag{5.16}
$$

or, since $t_2 = t_1$ for simultaneous events,

$$
L_0 = \gamma L \equiv \gamma (z_2 - z_1) \,.
\tag{5.17}
$$

Thus, the observer at O sees a *contracted* length L_0/γ. At first sight, this result seems to depend on which frame is which, contrary to the first postulate. But in fact, we could have chosen to measure the length in the laboratory frame and we could have then asked the observer at O' to determine the distance in the moving frame. In this case,

$$
L' = \gamma \left[L_0 - (t_2 - t_1)\, v \right] ,
\tag{5.18}
$$

where t_2 and t_1 are the different times that correspond to the one moment t' at which the observer at O' sees the end points z_1' and z_2':

$$
t' = \gamma \left(t_1 - \frac{v\, z_1}{c^2} \right) = \gamma \left(t_2 - \frac{v\, z_2}{c^2} \right) .
\tag{5.19}
$$

Rearranging terms, we get

$$
t_2 - t_1 = \frac{(z_2 - z_1)\, v}{c^2} = \frac{L_0\, v}{c^2} ,
\tag{5.20}
$$

so that

$$L' = \gamma L_0 \left(1 - \frac{v^2}{c^2}\right)$$

$$= \frac{L_0}{\gamma} . \qquad (5.21)$$

Both observers come to the same conclusion!

A similar phenomenon occurs when events are measured at different times, but in the same spatial location. Here, $z_1' = z_2' \equiv z'$ and $\Delta t' = t_2' - t_1'$. The observer at O, however, sees the events occurring at different values of z:

$$z_1 = \gamma (z' + v t_1')$$

$$\neq z_2 = \gamma (z' + v t_2') . \qquad (5.22)$$

Thus,

$$\Delta t \equiv t_2 - t_1 = \gamma \left[t_2' - t_1' + v \frac{z_2' - z_1'}{c^2} \right]$$

$$= \frac{\Delta t'}{\sqrt{1 - (v/c)^2}}$$

$$> \Delta t' . \qquad (5.23)$$

Evidently, an observer finds a dilatation of the time between two events taking place at a fixed point in a moving frame, compared to the time interval reported from the rest frame of that point.

Let us now broaden our exploration of the new framework, and consider additional consequences of the Lorentz transformation. Although spatial and time intervals need not be identical in frames moving relative to each other, *dimensionless* quantities, such as the number of events, must be preserved. (Remember, the physical laws must be invariant, so if a π° decays into two photons in one frame, it must also be seen to decay into two photons in every other frame.) For example, the phase of a wave must be invariant:

$$\Omega \equiv \omega t - \mathbf{k} \cdot \mathbf{x} = \omega' t' - \mathbf{k}' \cdot \mathbf{x}' , \qquad (5.24)$$

so that

$$\omega t - k z = \omega' \left(\gamma t - \gamma \frac{v z}{c^2} \right) - k' (\gamma z - \gamma v t) , \qquad (5.25)$$

or

$$\omega t - k z = (\omega' \gamma + \gamma k' v) t - \left(\gamma \frac{\omega' v}{c^2} + \gamma k' \right) z , \qquad (5.26)$$

for a wave propagating in the z-direction. Since this must be true for all values of t and z, we conclude that

$$\omega = \gamma \left(\omega' + k' v \right),\tag{5.27}$$

and

$$k = \gamma \left(k' + \frac{v \omega'}{c^2} \right).\tag{5.28}$$

For *light waves*, the frequencies and wavenumbers are simply related as $k = \omega/c$ and $k' = \omega'/c$, so

$$\omega = \gamma \omega' \left(1 + \beta \right),\tag{5.29}$$

or

$$\omega' = \gamma \omega \left(1 - \beta \right),\tag{5.30}$$

where $\beta \equiv v/c$. In general, the wavenumber **k** will not be directed along the z-axis, and this relation, known as the *Doppler shift formula*,[16] is then written as

$$\omega' = \gamma \omega \left(1 - \beta \cos\theta \right),\tag{5.31}$$

where $\mathbf{k} \cdot \mathbf{v} = \cos\theta \, |\mathbf{k}| \, |\mathbf{v}|$.

A striking feature of equations (5.27) and (5.28) is that the quantities $(\omega/c, \mathbf{k})$ *have the same (generalized) transformation properties* as $(ct, \mathbf{x})$. However, before we can proceed to investigate the *deeper meaning* of this similarity (see § 5.2 below), let us first consider another aspect of the basic Lorentz transformation and how it relates to the Galilean relativity we had been considering earlier. Let us return for the moment to the two frames moving with O and O', and let us suppose that a particle's motion is such that the observer at O detects a displacement $d\mathbf{x}\,(x, y, z)$ during the time interval dt. According to the observer at O', the corresponding intervals are $d\mathbf{x}'\,(x', y', z')$ and dt'. The transformation properties we have been discussing suggest that

$$\frac{dx'}{dt'} = \frac{dx}{\gamma \left(dt - v\, dz/c^2 \right)},\tag{5.32}$$

$$\frac{dy'}{dt'} = \frac{dy}{\gamma \left(dt - v\, dz/c^2 \right)},\tag{5.33}$$

and

$$\frac{dz'}{dt'} = \frac{dz - v\, dt}{\left(dt - v\, dz/c^2 \right)}.\tag{5.34}$$

---

[16]This expression predicts a *transverse* Doppler shift, which was verified by several early experiments, including one using the Mössbauer effect (Hay, Schiffer, Cranshaw, and Egelstaff 1960).

Expressed in terms of the velocities $\mathbf{u} = d\mathbf{x}/dt \equiv (u^1, u^2, u^3)$ and $\mathbf{u}' = d\mathbf{x}'/dt'$ measured within the respective frames, these expressions reduce to the simpler form

$$u'^1 = \frac{u^1}{\gamma\,(1 - v\,u^3/c^2)}\,, \tag{5.35}$$

$$u'^2 = \frac{u^2}{\gamma\,(1 - v\,u^3/c^2)}\,, \tag{5.36}$$

and

$$u'^3 = \frac{u^3 - v}{(1 - v\,u^3/c^2)}\,. \tag{5.37}$$

The reciprocal transformations are

$$u^1 = \frac{u'^1}{\gamma\,(1 + v\,u'^3/c^2)}\,, \tag{5.38}$$

$$u^2 = \frac{u'^2}{\gamma\,(1 + v\,u'^3/c^2)}\,, \tag{5.39}$$

and

$$u^3 = \frac{u'^3 + v}{(1 + v\,u'^3/c^2)}\,. \tag{5.40}$$

These relations, describing the relativistic addition of velocities, allow us to make the following two remarks. First, we note that when $v \ll c$, $\gamma \to 1$, and

$$\begin{aligned}
u^1 &= u'^1\,, \\
u^2 &= u'^2\,, \\
u^3 &= u'^3 + v\,,
\end{aligned} \tag{5.41}$$

which are just the results expected from Galilean relativity. Thus, the *special theory of relativity* that we are now considering reduces to the nonrelativistic limit when the velocity is small. Second, when the velocity is relativistic, i.e., $v \to c$, these transformations retain the validity of our second postulate, for in this limit, $u^3 \to (c + v)/(1 + v/c) = c$. That is, all observers still measure the same maximum velocity, c.

5.2 MATHEMATICAL STRUCTURE OF FOUR-DIMENSIONAL SPACETIME

The strikingly similar behavior of $(\omega/c, \mathbf{k})$ and $(ct, \mathbf{x})$ under a Lorentz transformation is an inkling of a more elaborate substructure in special relativity.[17] The four-dimensional (4D) quantities

$$x^\mu \equiv (x^0, \mathbf{x}) \qquad (\mu = 0, 1, 2, 3) , \tag{5.42}$$

and

$$k^\mu \equiv (k^0, \mathbf{k}) \qquad (\mu = 0, 1, 2, 3) , \tag{5.43}$$

where $x^0 \equiv ct$ and $k^0 \equiv \omega/c$, are examples of 4-vectors in the 4D spacetime. Our notation here is such that 4-vectors are labeled with a superscript, to distinguish them from the covectors to be defined below. More specifically, x^μ and k^μ *transform* as 4-vectors under a Lorentz transformation. Any quantity that transforms in the same fashion as x^μ is a 4-vector in this space.

The spacetime interval Δs is itself a special quantity because it is *invariant*:

$$
\begin{aligned}
(\Delta s)^2 &= (x^0)^2 - (x^1)^2 - (x^2)^2 - (x^3)^2 \\
&= \gamma^2 \left(x'^0 + \beta \frac{x'^3}{c} \right)^2 - (x^1)^2 - (x^2)^2 - \gamma^2 \left(x'^3 + \frac{v\,x'^0}{c} \right)^2 \\
&= (x'^0)^2 - (x'^1)^2 - (x'^2)^2 - (x'^3)^2 \\
&= (\Delta s')^2 .
\end{aligned}
\tag{5.44}
$$

Thus, s is a *scalar* in 4D spacetime. In the rest frame of an observer for whom $|\mathbf{x}_2 - \mathbf{x}_1| = 0$,

$$s^2 = c^2 \tau^2 , \tag{5.45}$$

where τ is the *proper* time, i.e., it is the time interval measured in a frame where the events that define this interval occur at the *same* spatial location. In a different frame,

$$
\begin{aligned}
s^2 &= (x^0)^2 \left(1 - \frac{|\mathbf{x}|^2}{(x^0)^2} \right) \\
&= c^2 t^2 (1 - \beta^2) \\
&= c^2 t^2 / \gamma^2 ,
\end{aligned}
\tag{5.46}
$$

---

[17]Two highly recommended books on this subject are Sard (1970) and the more elaborate text by Weinberg (1972).

which results in the expression $t = \gamma \tau$, as expected from our earlier discussion of time dilatation. When v is not constant, these intervals must be taken in their infinitesimal limits:

$$ds^2 = c\,dt^2 - (d\,x^1)^2 - (d\,x^2)^2 - (d\,x^3)^2 , \tag{5.47}$$

and

$$dt = \gamma\,d\tau , \tag{5.48}$$

for which the Lorentz invariant is then ds.

The mathematical properties of spacetime are those of the group of all transformations that leave s^2 (or ds^2) invariant. Another way of expressing our first postulate is that the laws of nature must be *invariant in form* under the transformations of the Lorentz group:

$$x'^\alpha = x'^\alpha (x^0, x^1, x^2, x^3) \qquad (\alpha = 0, 1, 2, 3) . \tag{5.49}$$

Tensors of rank k associated with the spacetime point x are *defined* by their transformation properties under the transformation

$$x \to x' . \tag{5.50}$$

The three quantities x^μ, k^μ, and s that we have encountered thus far are (special) examples of these tensors, which we now define:

(i) A *scalar* (i.e., a tensor of rank 0) is a single function of x whose value is *not* changed by the transformation. The Lorentz interval s^2 is a scalar.

(ii) A *vector* (i.e., a tensor of rank 1) is a set of ordered numbers that transform according to the rule

$$V'^\alpha = \frac{\partial x'^\alpha}{\partial x^\beta} V^\beta \tag{5.51}$$

(a *"contravariant"* vector), or

$$V'_\alpha = \frac{\partial x^\beta}{\partial x'^\alpha} V_\beta \tag{5.52}$$

(a *"covariant"* vector). Throughout this book, unless otherwise noted, a repeated index means that the term in which the index appears is to be summed over all its values. So, for example, $x^\alpha x_\alpha \equiv x^0 x_0 + x^1 x_1 + x^2 x_2 + x^3 x_3$. Note that differentiation with respect to a contravariant component of the coordinate vector transforms as the component of a covariant 4-vector operator:

$$\frac{\partial}{\partial x'^\alpha} = \frac{\partial x^\beta}{\partial x'^\alpha} \frac{\partial}{\partial x^\beta} . \tag{5.53}$$

For the Lorentz transformation we considered earlier, the transformation coefficients

$$\alpha^{\alpha}{}_{\beta} \equiv \frac{\partial x'^{\alpha}}{\partial x^{\beta}} \, , \tag{5.54}$$

such that

$$V'^{\alpha} = \alpha^{\alpha}{}_{\beta} \, V^{\beta} \, , \tag{5.55}$$

are given explicitly as

$$\begin{aligned}
\alpha^{1}{}_{1} &= +1 \, , & \alpha^{2}{}_{2} &= +1 \, , \\
\alpha^{3}{}_{3} &= \gamma \, , & \alpha^{3}{}_{0} &= -v\,\gamma/c \, , \\
\alpha^{0}{}_{0} &= \gamma \, , & \alpha^{0}{}_{3} &= -v\,\gamma/c \, ,
\end{aligned} \tag{5.56}$$

and

$$\begin{aligned}
\alpha_{1}{}^{1} &= +1 \, , & \alpha_{2}{}^{2} &= +1 \, , \\
\alpha_{3}{}^{3} &= \gamma \, , & \alpha_{3}{}^{0} &= v\,\gamma/c \, , \\
\alpha_{0}{}^{0} &= \gamma \, , & \alpha_{0}{}^{3} &= v\,\gamma/c \, ,
\end{aligned} \tag{5.57}$$

with all others zero. Notationally, it is important to remember that the indices in these coefficients may be either subscripts or superscripts and their horizontal ordering is not arbitrary. A superscript means that its corresponding coordinate is being differentiated (in Equations [5.51] and [5.52]). The first horizontal position is reserved for the "primed" coordinate. So, for example, $\alpha_{\mu}{}^{\nu} = \partial x^{\nu}/\partial x'^{\mu}$.

(iii) *Tensors* of rank $k > 1$ are defined by means of an obvious generalization of scalars and vectors. A *contravariant* tensor of rank 2, $T^{\alpha\beta}$, consists of 16 numbers that transform according to

$$T'^{\alpha\beta} = \alpha^{\alpha}{}_{\gamma} \, \alpha^{\beta}{}_{\delta} \, T^{\gamma\delta} \, . \tag{5.58}$$

A *covariant* tensor of rank 2 transforms according to

$$T'_{\alpha\beta} = \alpha_{\alpha}{}^{\gamma} \, \alpha_{\beta}{}^{\delta} \, T_{\gamma\delta} \, , \tag{5.59}$$

where

$$\alpha_{\alpha}{}^{\gamma} \equiv \frac{\partial x^{\gamma}}{\partial x'^{\alpha}} \, . \tag{5.60}$$

The transformation coefficients $\alpha^{\alpha}{}_{\beta}$ and $\alpha_{\alpha}{}^{\beta}$ have the very useful property that

$$\alpha^{\alpha}{}_{\beta} \, \alpha_{\alpha}{}^{\gamma} = \delta_{\beta}{}^{\gamma} \, , \tag{5.61}$$

where δ is the Kronecker delta (with $\delta_\alpha{}^\beta = 0$ for $\alpha \neq \beta$ and $\delta_\alpha{}^\alpha = 1$ for $\alpha = 0, 1, 2, 3$), as may be verified by the direct application of the chain rule of differentiation:

$$\frac{\partial x'^\alpha}{\partial x^\beta} \frac{\partial x^\gamma}{\partial x'^\alpha} = \frac{\partial x^\gamma}{\partial x^\beta} = \delta_\beta{}^\gamma \ . \tag{5.62}$$

Several important relations follow immediately. First, $V^\alpha W_\alpha$ is a scalar, since

$$\begin{aligned} V'^\alpha W'_\alpha &= \alpha^\alpha{}_\beta V^\beta \alpha_\alpha{}^\gamma W_\gamma \\ &= \delta_\beta{}^\gamma V^\beta W_\gamma \\ &= V^\beta W_\beta \ . \end{aligned} \tag{5.63}$$

Second, $S_{\alpha\beta} \equiv V_\alpha W_\beta$ is a tensor of rank 2, since

$$\begin{aligned} S'_{\alpha\beta} &= V'_\alpha W'_\beta \\ &= \alpha_\alpha{}^\gamma V_\gamma \alpha_\beta{}^\delta W_\delta \\ &= \alpha_\alpha{}^\gamma \alpha_\beta{}^\delta S_{\gamma\delta} \ . \end{aligned} \tag{5.64}$$

It is equally straightforward to show that $T_\alpha{}^\alpha$ is a scalar, and that $W_\alpha \equiv T_\alpha{}^\beta V_\beta$ is a covector. We especially note that in order for ds to be a scalar, it must be the product of contravariant and covariant vectors (see Equation [5.63]). Indeed, we may write

$$(ds)^2 = (dx^0)^2 - (dx^1)^2 - (dx^2)^2 - (dx^3)^2 \tag{5.65}$$

as a special case of the differential element

$$(ds)^2 = g_{\alpha\beta} \, dx^\alpha \, dx^\beta \ , \tag{5.66}$$

where $g_{\alpha\beta} = g_{\beta\alpha}$ is the *metric tensor*. In special relativity (as opposed to the general theory), $g_{\alpha\beta}$ is diagonal when using some special coordinate systems, such as Cartesian, for which

$$\begin{aligned} g_{00} &= +1 \\ g_{11} &= g_{22} = g_{33} = -1 \ . \end{aligned} \tag{5.67}$$

The contravariant metric tensor $g^{\alpha\beta}$ is defined to be the inverse of $g_{\alpha\beta}$, such that $x^\alpha = g^{\alpha\beta}(g_{\beta\gamma}x^\gamma)$, with $g^{\alpha\beta} g_{\beta\gamma} = \delta^\alpha{}_\gamma$. In special relativity their corresponding coefficients are the same when written in Cartesian coordinates:

$$g^{\alpha\beta} = g_{\alpha\beta} \ . \tag{5.68}$$

It is clear then, that

$$(ds)^2 = dx_\beta \, dx^\beta \ , \tag{5.69}$$

where

$$dx_\beta = g_{\beta\alpha}\, dx^\alpha \,, \tag{5.70}$$

a direct consequence of the definition of $(ds)^2$ (Equation [5.65]) and the values of $g_{\alpha\beta}$ in special relativity. "Contraction" with $g_{\alpha\beta}$ or $g^{\alpha\beta}$ is in fact the general procedure for changing an index on any tensor from contravariant to covariant, and vice versa.

As a final example, we will use these ideas and definitions to prove that the quantity $\partial\Phi/\partial x^0 + \vec{\nabla}\cdot\mathbf{A}$ is a scalar and that Φ and $\mathbf{A}$ must therefore form a 4-vector potential $A^\alpha \equiv (\Phi, \mathbf{A})$. In Cartesian coordinates, we can write

$$
\begin{aligned}
\frac{\partial\Phi'}{\partial x'^0} + \vec{\nabla}'\cdot\mathbf{A}' &\equiv \frac{\partial A'^\sigma(x')}{\partial x'^\sigma} \\[2mm]
&= \frac{\partial}{\partial x'^\sigma}\left[\alpha^\sigma{}_\beta\, A^\beta(x)\right] \\[2mm]
&= \alpha^\sigma{}_\beta\, \frac{\partial A^\beta(x)}{\partial x'^\sigma} \\[2mm]
&= \alpha^\sigma{}_\beta\, \frac{\partial A^\beta(x)}{\partial x^\alpha}\frac{\partial x^\alpha}{\partial x'^\sigma} \\[2mm]
&= \alpha^\sigma{}_\beta\, \alpha_\sigma{}^\alpha\, \frac{\partial A^\beta(x)}{\partial x^\alpha} \\[2mm]
&= \delta_\beta{}^\alpha\, \frac{\partial A^\beta(x)}{\partial x^\alpha} \\[2mm]
&= \frac{\partial\Phi}{\partial x^0} + \vec{\nabla}\cdot\mathbf{A} \,. \tag{5.71}
\end{aligned}
$$

Since $\partial/\partial x^\alpha$ transforms as a 4-vector and the left-hand side of Equation (5.71) is a scalar, then by Equation (5.63), the potential A^α must itself be a 4-vector in this spacetime.

5.3   LORENTZ TRANSFORMATION PROPERTIES OF PHYSICAL QUANTITIES

Mass represents a particle's *inertia* to acceleration. It is not surprising, therefore, that in special relativity different observers attribute different "masses" to the same object since velocities and accelerations do not, in general, add as they do in Galilean relativity. How, then, does one calculate the force on a moving particle? As we shall see, this is but one instance where our cleanest approach is to carry out our analysis of the interactions in specially selected frames first. As you can imagine, the special frame is that in which the particle is at *rest*. The *rest mass*, representing the particle's inertia in its own frame, is a quantity upon which all observers can agree, and we shall therefore always mean this particular quantity when we refer to the "mass." Our expectation is that the variation in the particle's inertia from frame to frame can therefore be represented in other ways.

The rest frame also happens to be that in which we are justified in using (the classically derived) Newton's laws since the particle's velocity v (which is in fact 0) is trivially much smaller than the speed of light, c (see § 5.1 above). The idea is to calculate the applied force in this frame—which we already know how to do from classical mechanics—and then to transform the result to any other frame by using an appropriate Lorentz transformation on the 4-vector representing that force, constructed from other scalars, 4-vectors, and 4-tensors using the techniques described in the previous subsection.

To begin with, let us *define* the 4-vector

$$f^\alpha \equiv m \frac{d^2 x^\alpha}{d\tau^2} ,$$ (5.72)

which we will call the *relativistic force*, and then examine its properties vis-à-vis the usual Newtonian force $\mathbf{F} = m\mathbf{a}$. In the particle's rest frame, where $d\tau = dt$, we have

$$\begin{aligned} f^i &= F^i &(i = 1, 2, 3) \\ f^0 &= 0 , \end{aligned}$$ (5.73)

with F^i the Cartesian components of $\mathbf{F}$. This 4-vector f^α looks like it might provide what we need, but what happens to it under a transformation? We know that

$$d x'^\alpha = \alpha^\alpha{}_\beta \, dx^\beta ,$$ (5.74)

and that $d\tau$ is invariant. Thus,

$$f'^\alpha = \alpha^\alpha{}_\beta \, f^\beta ,$$ (5.75)

as expected of a 4-vector. Substituting for f^α from equation (5.73), we immediately get

$$\begin{aligned} f_L^0 &= \frac{\gamma}{c} \mathbf{v} \cdot \mathbf{F} \\ \mathbf{f}_L &= \mathbf{F} + (\gamma - 1) \mathbf{v} \frac{\mathbf{v} \cdot \mathbf{F}}{v^2} , \end{aligned}$$ (5.76)

where f_L^α is the "4-force" in the frame moving with velocity $\mathbf{v}$ relative to the particle (i.e., the lab frame). Evidently, f_L^0 is proportional to the power exerted by the force on the particle.

This relativistic expression for the force suggests that we *define* an energy-momentum 4-vector

$$p^\alpha = m \frac{d\,x^\alpha}{d\,\tau} \,, \tag{5.77}$$

so that Newton's second law can be cast in the form

$$\frac{d\,p^\alpha}{d\,\tau} = f^\alpha \,. \tag{5.78}$$

Since $d\tau = dt/\gamma$, we see that

$$
\begin{aligned}
p^0 &= \gamma\,mc \\
\text{and} \quad \mathbf{p} &= \gamma\,m\,\mathbf{v} \,,
\end{aligned}
\tag{5.79}
$$

so the vector component, $\mathbf{p}$, of p^α has the "correct" limiting form when $\gamma \to 1$. The modification introduced by special relativity is due entirely to the inertia, which now varies with particle speed as discussed above. But what do we make of p^0? Since p^α is a product of a 4-vector and a scalar, it too is a 4-vector and the contraction

$$p^\alpha\,p_\alpha = (p^0)^2 - |\mathbf{p}|^2 \tag{5.80}$$

must be invariant. That is,

$$
\begin{aligned}
(p^0)^2 - |\mathbf{p}|^2 &= \gamma^2\,m^2\,c^2 - m^2\,\gamma^2\,|\mathbf{v}|^2 \\
&= m^2\,c^2 = \text{constant} \,,
\end{aligned}
\tag{5.81}
$$

which directly couples p^0 to $\mathbf{p}$ in all frames! More specifically, in the limit $v \to 0$,

$$
\begin{aligned}
\gamma\,m\,c^2 &= \frac{m\,c^2}{\sqrt{1 - (v/c)^2}} \\
&\approx m\,c^2 \left[1 + \frac{1}{2} \left(\frac{v}{c} \right)^2 + \frac{3}{8} \left(\frac{v}{c} \right)^4 + \dots \right] \\
&\approx m\,c^2 + \frac{1}{2}\,m\,v^2 \,.
\end{aligned}
\tag{5.82}
$$

This term is therefore just the sum of the kinetic energy and another form of energy that does not vanish even when $v = 0$. Not surprisingly, we refer to it as the *particle's rest energy*, and we refer to the sum

$$E \equiv \gamma\,m\,c^2 \tag{5.83}$$

as its *relativistic kinetic energy*. It is now clear that

$$p^0 = E/c \, . \tag{5.84}$$

These definitions of energy and momentum lead to several other useful relations. For example, the particle's velocity may be expressed in terms of $\mathbf{p}$ and E:

$$\mathbf{v} = \frac{\mathbf{p}}{\gamma m} = \frac{c^2 \mathbf{p}}{E} \, , \tag{5.85}$$

which gives

$$E_\nu = p_\nu \, c \tag{5.86}$$

for a photon. And combining expressions (5.80) and (5.84), we obtain one of the most celebrated equations of special relativity, viz.,

$$\boxed{E^2 = |\mathbf{p}|^2 \, c^2 + m^2 \, c^4} \, . \tag{5.87}$$

To complete this discussion, we must still examine the consequences of the requirement that the total vector momentum

$$\mathbf{P} = \sum_i \mathbf{p}_i \tag{5.88}$$

and the total energy

$$E_{\text{tot}} = \sum_i E_i \tag{5.89}$$

for different particle species i be *conserved* in an isolated system. By the first postulate of special relativity—that the laws of physics should be invariant under a Lorentz transformation—the relationship between $\mathbf{P}$ and E_{tot} must be preserved from frame to frame. Thus, the set of numbers $P^\alpha \equiv (P^0, \mathbf{P})$, where

$$P^0 = \frac{E_{\text{tot}}}{c}$$

$$\text{and} \quad \mathbf{P} = \sum_i \mathbf{p}_i \, , \tag{5.90}$$

must be a 4-vector in 4D spacetime. For example, let us consider a two-particle interaction, with initial and final 4-momenta $(p_1{}^\mu, p_2{}^\mu)$ and $(p_3{}^\mu, p_4{}^\mu)$, respectively (Figure 5.4). Then,

$$P^\mu = p_1{}^\mu + p_2{}^\mu = p_3{}^\mu + p_4{}^\mu \, , \tag{5.91}$$

and

$$
\begin{aligned}
P^\mu P_\mu &= (m^*)^2 c^2 \\
&= p_1{}^\mu p_{1\mu} + p_2{}^\mu p_{2\mu} + 2p_1{}^\mu p_{2\mu} \\
&= p_3{}^\mu p_{3\mu} + p_4{}^\mu p_{4\mu} + 2p_3{}^\mu p_{4\mu} ,
\end{aligned}
\tag{5.92}
$$

where m^* is the composite mass in the center of mass frame, i.e., the frame in which $\mathbf{P} = 0$. Suppose m_2 is initially at rest. Then,

$$
p_2{}^\mu = (m_2\, c, 0) ,
\tag{5.93}
$$

and

$$
\begin{aligned}
(m^*)^2 c^2 &= m_1^2 c^2 + m_2^2 c^2 + 2m_1\, m_2\, c^2\, \gamma_1 \\
\text{or} \qquad m^* &= \sqrt{m_1^2 + m_2^2 + 2m_1\, m_2\, \gamma_1} \ .
\end{aligned}
\tag{5.94}
$$

Notice that the mass in this frame, i.e., the energy available for particle creation, goes only as the square root of the beam energy, $\gamma_1\, m_1\, c^2$. This is the reason, incidentally, why ideal particle accelerators have colliding beams in the laboratory frame (which then becomes the center of mass frame!) to make more efficient use of the acceleration.

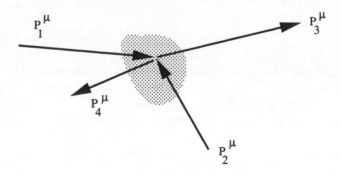

Figure 5.4   A two-particle collision, with entering channels 1 and 2 and exit channels 3 and 4.

In the final state of this two-particle interaction,

$$
(m^*)^2 c^2 = m_3^2 c^2 + m_4^2 c^2 + 2m_3\, m_4\, c^2\, \gamma_2\, \gamma_4 - 2\,\mathbf{P}_3 \cdot \mathbf{P}_4 ,
\tag{5.95}
$$

where as usual, $\mathbf{p}_i = \gamma_i \, m_i \, \mathbf{v}_i$. Thus, if $m_3 = m_1$ and $m_4 = m_2$, we get

$$2m_1 \, m_2 \, c^2 \, \gamma_1 - 2m_3 \, m_4 \, c^2 \, \gamma_3 \, \gamma_4 = -2\mathbf{p}_3 \cdot \mathbf{p}_4 \, , \qquad (5.96)$$

or

$$\mathbf{p}_3 \cdot \mathbf{p}_4 = m_1 \, m_2 \, c^2 \, (\gamma_3 \, \gamma_4 - \gamma_1) \, , \qquad (5.97)$$

which relates the angle between the final momentum states to the incoming particle energies.

Of course, much, much more can be gleaned from the conservation of P^μ, but this would take us beyond our present scope.[18] We will now stop here and turn our attention to bringing the theory of electrodynamics into the fold of this new relativistic framework.

5.4 LORENTZ TRANSFORMATION OF MACROSCOPIC ELECTRODYNAMICS

In order for the theory of electrodynamics to be consistent with the principles of special relativity, the equations used to describe its physical laws must be equally applicable in every equivalent reference frame. Thus, to make electrodynamics Lorentz invariant, we need to reformulate our equations in terms of 4-vectors and other 4D covariants.

As we have emphasized all along, the basic physical laws are expressions for the *field increments* generated by *conserved* charges. This conservation of charge is the basis of the continuity equation in a given inertial frame (Equation [1.32]). But in order for our theory to be Lorentz invariant, this alone is not sufficient— we need the charge to be conserved from frame to frame as well, i.e., *we need the charge to be a Lorentz scalar*. To see whether or not charge as we have defined it fulfills this pivotal role, let's begin by writing the charge conservation Equation (1.32) in 4D language, i.e.,

$$\frac{\partial J^\alpha}{\partial x^\alpha} \equiv \partial_\alpha J^\alpha = 0 \, , \qquad (5.98)$$

where

$$J^\alpha \equiv (c\rho, \mathbf{J}) \, . \qquad (5.99)$$

----

[18]The reference of choice for most matters dealing with relativistic particle kinematics is Byckling and Kajantie (1973).

Since Equation (1.32) is satisfied in every inertial reference frame, so too is Equation (5.98). Evidently, the quantity $\partial_\alpha J^\alpha$ is a Lorentz scalar, and since ∂_α transforms as a covariant 4-vector (see Equation [5.53]), we know by the result of (5.63) that J^α must therefore be a *contravariant* 4-vector.

Now, whenever any current J^α satisfies the invariant conservation law (5.98), it is possible to form a total charge

$$Q \equiv \frac{1}{c} \int d^3x \, J^0(\mathbf{x}) \,, \tag{5.100}$$

which is time independent because

$$
\begin{aligned}
\frac{dQ}{dt} &= \int d^3x \, \frac{\partial J^0}{\partial x^0} \\
&= -\int d^3x \, \vec{\nabla} \cdot \mathbf{J} \\
&= 0 \,.
\end{aligned}
\tag{5.101}
$$

The second line in this equation is the result of the conservation law (5.98), and the last equality follows from the application of Gauss's theorem under the assumption that $\mathbf{J}(\mathbf{x}) \to 0$ as $|\mathbf{x}| \to \infty$.

So far, we have done nothing more than confirm the derivation of Equation (1.32). But let us next rewrite Equation (5.100) as

$$Q = \int d^4x \, J^\alpha(x) \, \partial_\alpha \Theta(\eta_\beta x^\beta) \,, \tag{5.102}$$

where Θ is the Heaviside (step) function

$$
\Theta(u) = \begin{cases} 1 & \text{if } u > 0 \\ 0 & \text{if } u \leq 0 \,, \end{cases}
\tag{5.103}
$$

and η_μ is defined by

$$\eta_1 \equiv \eta_2 \equiv \eta_3 \equiv 0, \qquad \eta_0 = +1 \,. \tag{5.104}$$

To see that Equations (5.100) and (5.102) are equivalent, note that $\partial_\alpha \Theta(\eta_\beta x^\beta)$ $= \delta(\eta_\beta x^\beta) \, \delta_\alpha{}^\mu \, \eta_\mu$ so that Equation (5.100) results from (5.102) with just a single integration over dx^0, which eliminates the Dirac delta function $\delta(\eta_\beta x^\beta)$. Under a Lorentz transformation, we get $Q \to Q'$, where

$$Q' = \int d^4x \, \alpha^\alpha{}_\lambda \, J^\lambda \, \alpha_\alpha{}^\mu \, \partial_\mu \, \Theta(\eta_\beta \, \alpha^\beta{}_\delta \, x^\delta) \,, \tag{5.105}$$

which by Equation (5.61) may be written

$$Q' = \int d^4x \, J^\alpha(x) \, \partial_\alpha \Theta(\eta'_\beta \, x^\beta) \,, \qquad (5.106)$$

where $\eta'_\beta \equiv \alpha^\nu{}_\mu \, \eta_\nu$. Thus, the effect on Q is apparently just to change η_μ, and using (5.98), we can write the difference in Q as

$$Q - Q' = \int d^4x \, \partial_\alpha \left\{ J^\alpha(x) \left[\Theta(\eta'_\beta \, x^\beta) - \Theta(\eta_\beta \, x^\beta) \right] \right\} \,. \qquad (5.107)$$

This provides us with the desired result because $J^\alpha(x) \to 0$ as $|\mathbf{x}| \to \infty$ with t fixed, whereas the function $\Theta(\eta'_\beta \, x^\beta) - \Theta(\eta_\beta \, x^\beta)$ vanishes when $t \to \infty$ with $\mathbf{x}$ fixed, so that the 4D generalization of Gauss's theorem immediately gives

$$Q - Q' = 0 \,. \qquad (5.108)$$

We have thus proved the very important result that Q is not only conserved in any given inertial frame, but that it is also a Lorentz scalar, ensuring that it has the *same* value in all inertial frames.

Thus assured, we can now confidently proceed to reformulate the equations governing the electric and magnetic fields in terms of Lorentz scalars, 4-vectors, and 4-tensors. A reasonable starting point is the 4D potential $A^\alpha \equiv (\Phi, \mathbf{A})$, which was shown to be a 4-vector in § 5.2. above. According to Equations (1.35) and (1.37),

$$\mathbf{E} = -\frac{1}{c} \frac{\partial \mathbf{A}}{\partial t} - \vec{\nabla}\Phi \,, \qquad (5.109)$$

and

$$\mathbf{B} = \vec{\nabla} \times \mathbf{A} \,. \qquad (5.110)$$

Therefore,

$$
\begin{aligned}
E^i &= -\frac{1}{c}\frac{\partial A^i}{\partial t} - \frac{\partial \Phi}{\partial(-x_i)} \\
&= -\frac{\partial A^i}{\partial x^0} - \frac{\partial A^0}{\partial(-x_i)} \\
&= -\frac{\partial A^i}{\partial x_0} + \frac{\partial A^0}{\partial x_i} \\
&= -\left(\partial^0 A^i - \partial^i A^0\right) \,.
\end{aligned}
\qquad (5.111)
$$

With a similar derivation, we find that

$$
\begin{aligned}
B_i &= -\varepsilon_{ijk} \frac{\partial A^j}{\partial(-x_k)} \\
&= \varepsilon_{ijk} \, \partial^k A^j \,,
\end{aligned}
\qquad (5.112)
$$

where ε_{ijk} is defined by

$$\varepsilon = \begin{cases} +1 & \text{if } ijk \text{ forms an even permutation of 123} \\ -1 & \text{if } ijk \text{ forms an odd permutation of 123} \\ 0 & \text{otherwise .} \end{cases} \qquad (5.113)$$

These equations thus suggest that $\mathbf{E}$ and $\mathbf{B}$ have components that are elements of a second-rank, antisymmetric *field-strength tensor:*

$$\boxed{F^{\alpha\beta} = \partial^\alpha A^\beta - \partial^\beta A^\alpha} \ . \qquad (5.114)$$

Written out explicitly, we have

$$F^{\alpha\beta} = \begin{pmatrix} 0 & -E^x & -E^y & -E^z \\ E^x & 0 & -B^z & B^y \\ E^y & B^z & 0 & -B^x \\ E^z & -B^y & B^x & 0 \end{pmatrix} \ . \qquad (5.115)$$

This field-tensor knits the two 3-vector fields $\mathbf{E}$ and $\mathbf{B}$ into a single entity, resolvable on the four dimensions of spacetime in ways that convert electric into magnetic fields, and vice versa, merely by viewing the *electromagnetic field* from relatively moving frames. Its primary usefulness will be to permit us to cast the Maxwell equations into an explicitly covariant form.

First, the inhomogeneous equations are (1.12),

$$\vec{\nabla} \cdot \mathbf{E} = 4\pi\rho \ , \qquad (5.116)$$

which becomes

$$\partial_i F^{i0} = \frac{4\pi}{c} J^0 \ , \qquad (5.117)$$

and (1.34)

$$\vec{\nabla} \times \mathbf{B} - \frac{1}{c}\frac{\partial \mathbf{E}}{\partial t} = \frac{4\pi}{c} \mathbf{J} \ . \qquad (5.118)$$

These can be written together in covariant form as

$$\boxed{\partial_\alpha F^{\alpha\beta} = 4\pi J^\beta / c} \ , \qquad (5.119)$$

which may be easily verified with the definitions of $F^{\alpha\beta}$ and J^{β}. The decomposition of Equation (5.119) into its different components may be made with equal validity in any frame, consistent with the relativity principle.

Next, the homogeneous equations are (1.14),

$$\vec{\nabla} \cdot \mathbf{B} = 0 , \tag{5.120}$$

which becomes

$$\partial^1 F^{32} + \partial^2 F^{13} + \partial^3 F^{21} = 0 , \tag{5.121}$$

and (1.22)

$$\vec{\nabla} \times \mathbf{E} + \frac{1}{c} \frac{\partial \mathbf{B}}{\partial t} = 0 . \tag{5.122}$$

Both of these equations may be combined into the single form

$$\boxed{\partial^\alpha F^{\beta\gamma} + \partial^\beta F^{\gamma\alpha} + \partial^\gamma F^{\alpha\beta} = 0} . \tag{5.123}$$

To complete the set of electromagnetic equations, we also need to consider the Lorentz force law:

$$\frac{d\mathbf{p}}{dt} = q \left[\mathbf{E} + \frac{\mathbf{v}}{c} \times \mathbf{B} \right] . \tag{5.124}$$

In terms of the proper time interval $d\tau$ and the 4-velocity

$$u^\alpha \equiv \frac{dx^\alpha}{d\tau} = (\gamma c , \gamma \mathbf{v}) , \tag{5.125}$$

this becomes

$$\frac{d\mathbf{p}}{d\tau} = \frac{q}{c}(u^0 \mathbf{E} + \mathbf{u} \times \mathbf{B}) . \tag{5.126}$$

As we have already seen (Equation [5.76]), the quantity

$$\frac{dp^0}{d\tau} = f^0 \tag{5.127}$$

is the rate of change of the particle's energy (i.e., the power exerted on it), which we can also rewrite in terms of $F^{\alpha\beta}$:

$$\begin{aligned} f^0 &= \frac{\gamma}{c} \mathbf{v} \cdot \mathbf{F} \\ &= \frac{\gamma}{c} \mathbf{v} \cdot q \mathbf{E} \\ &= \frac{q}{c} F^0{}_i u^i . \end{aligned} \tag{5.128}$$

This covariant form generalizes to the other three indices, and since $dp^\alpha/d\tau$ is a 4-vector, the right-hand sides of Equations (5.126) and (5.128) constitute a representation of the covariant electromagnetic force on a charge q:

$$\boxed{f^\alpha \equiv dp^\alpha/d\tau = q\,F^\alpha{}_\gamma\,u^\gamma/c}\ . \tag{5.129}$$

At this point, it is appropriate to ask how the electromagnetic field behaves under a Lorentz transformation. Using our prescription for transforming contravariant tensors (see § 5.2 above), we find that

$$F'^{\alpha\beta} = \frac{\partial\,x'^\alpha}{\partial\,x^\gamma}\,\frac{\partial\,x'^\beta}{\partial\,x^\mu}\,F^{\gamma\mu}\ . \tag{5.130}$$

Thus, for the specific example of a Lorentz boost in the x^3 direction (whose coefficients are given by Equations [5.56] and [5.57]), we get

$$\begin{aligned}
E'^1 &= \gamma\,(E^1 - \beta\,B^2)\,,\\
E'^2 &= \gamma\,(E^2 + \beta\,B^1)\,,\\
E'^3 &= E^3\,,
\end{aligned} \tag{5.131}$$

and

$$\begin{aligned}
B'^1 &= \gamma\,(B^1 + \beta\,E^2)\,,\\
B'^2 &= \gamma\,(B^2 - \beta\,E^1)\,,\\
B'^3 &= B^3\ .
\end{aligned} \tag{5.132}$$

Ponder, if you will, the power of these equations, for they are truly monuments in the theory of electrodynamics. We began with definitions of the electric and magnetic fields in terms of the force experienced by a moving test charge in the presence of another charge, and here, after realizing that our initial formulation was not invariant under the classical relativity theory, and after we were given an inkling of a synthesis of the two fields through their representation in a single antisymmetric tensor $F^{\alpha\beta}$, we finally come to the appreciation of the fact that the concept of a pure electric or a pure magnetic field is not consistent with Lorentz invariance. It is precisely for this reason that we refer to it as the *electromagnetic field*—one quantity, unified in the special theory of relativity.

5.5 STRESS-ENERGY MOMENTUM TENSOR AND CONSERVATION LAWS

It was noted in Chapter 1 that the basic idea of electrodynamics as a field theory is that charges and currents produce at each point of spacetime a field that has a reality of its own. Thus, since ρ and $\mathbf{J}$ (or in covariant notation, the source J^α) are known to satisfy the continuity equation (5.98), it is not unreasonable to expect that the fields themselves should satisfy some type of conservation and/or transport equation. After all, the electromagnetic field is a dynamical entity possessing energy, momentum, and angular momentum.

The fields $\mathbf{E}$ and $\mathbf{B}$ are *operationally* defined by the force per unit of nonstatic test charge q. As we have seen, the covariant electromagnetic force on q is

$$f^\alpha = \frac{q}{c} F^\alpha{}_\mu u^\mu \, , \tag{5.133}$$

where

$$\mathbf{f} = \frac{d\mathbf{p}}{d\tau} \, , \tag{5.134}$$

and where

$$f^0 = \frac{\gamma}{c} \mathbf{v} \cdot q\mathbf{E} \tag{5.135}$$

is the rate of work done by the electromagnetic field on the source distribution. Thus, the 0th component of f^α should give us the *energy* of the field and its *flux density* via the energy transfer between the matter and the radiation. It is also expected that $\mathbf{f}$ will allow us to identify the *field momentum* and its flux through the momentum transfer between the particles and fields (i.e., the force). To carry out this identification, as we did, for example, to derive the Poynting vector and the (nonrelativistic) momentum flux, we need to eliminate the currents and densities from the equations and express them in terms of the field quantities only.

Let us write

$$f^\alpha = F^\alpha{}_\mu \left\{ \frac{q\,u^\mu}{c} \right\} \, . \tag{5.136}$$

Remembering that $u^\mu = (\gamma c, \gamma\mathbf{v})$, we see that

$$q\,u^\mu = (\gamma\,q\,c, \gamma\,q\,\mathbf{v}) \, . \tag{5.137}$$

But

$$\gamma\,q = \gamma\,\delta V\,\rho \, , \tag{5.138}$$

where $\gamma \, \delta V$ is the invariant volume. Thus,

$$\Phi^\alpha \equiv \frac{f^\alpha}{\gamma \, \delta V} = \frac{1}{c} F^\alpha{}_\mu J^\mu \qquad (5.139)$$

is the 4-vector *force density* due to the electromagnetic field.

We also know from the covariant form of Maxwell's equations that

$$\partial_\alpha F^{\alpha\beta} = \frac{4\pi}{c} J^\beta \; , \qquad (5.140)$$

and therefore Φ^α can be expressed entirely in terms of the field tensor:

$$\Phi^\alpha = F^\alpha{}_\mu \frac{\partial_\beta F^{\beta\mu}}{4\pi} \; . \qquad (5.141)$$

However, before we can identify the energy and momentum of the field, we still need to express this equation as a continuity equation or an expression of the conservation of "something." Let's put

$$4\pi \, \Phi^\alpha = \partial_\beta (F^{\beta\mu} F^\alpha{}_\mu) - F^{\beta\mu} \, \partial_\beta (F^\alpha{}_\mu) \; . \qquad (5.142)$$

Now, using the metric tensor $g^{\alpha\beta}$ defined in Equation (5.67), we have

$$\begin{aligned}
F^{\beta\mu} \, \partial_\beta F^\alpha{}_\mu &= g^{\alpha\nu} F^{\beta\mu} \, \partial_\beta F_{\nu\mu} \\
&= g^{\alpha\nu} F^{\mu\beta} \, \partial_\mu F_{\nu\beta} \\
&= g^{\alpha\nu} F^{\beta\mu} \, \partial_\mu F_{\beta\nu} \qquad (5.143)
\end{aligned}$$

(with an interchange of the "dummy" indices μ and β in the second step and the use of the antisymmetric property of $F^{\mu\beta}$ in the last). But using the second Maxwell equation (5.123), we can also write this as

$$F^{\beta\mu} \, \partial_\beta F^\alpha{}_\mu = -F^{\beta\mu} \left\{ \partial^\alpha F_{\mu\beta} + \partial_\mu F_\beta{}^\alpha \right\} \; . \qquad (5.144)$$

Adding Equations (5.143) and (5.144) gives

$$F^{\beta\mu} \, \partial_\beta F^\alpha{}_\mu = -\frac{1}{2} F^{\beta\mu} \, \partial_\nu F_{\mu\beta} \, g^{\alpha\nu} \; , \qquad (5.145)$$

so that using the antisymmetric property of $F^{\alpha\beta}$ and substituting λ for the index β, we get

$$\begin{aligned}
F^{\beta\mu} \, \partial_\beta F^\alpha{}_\mu &= -\frac{1}{4} \partial_\nu \left(g^{\alpha\nu} F^{\beta\mu} F_{\mu\beta} \right) \\
&= +\frac{1}{4} \partial_\nu \left(g^{\alpha\nu} F^{\lambda\mu} F_{\lambda\mu} \right) \; . \qquad (5.146)
\end{aligned}$$

Thus, we can write

$$
\begin{aligned}
4\pi\Phi^\alpha &= \partial_\beta\left(F^{\beta\mu}\,F^\alpha{}_\mu\right) - \frac{1}{4}\,\partial_\nu\left(g^{\alpha\nu}\,F^{\lambda\mu}\,F_{\lambda\mu}\right) \\
&= \partial_\beta\left(F^{\beta\mu}\,F^\alpha{}_\mu - \frac{1}{4}\,\delta^\beta{}_\nu\,g^{\alpha\nu}\,F^{\lambda\mu}\,F_{\lambda\mu}\right) \\
&= \partial_\beta\left(F^{\beta\mu}\,F^\alpha{}_\mu - \frac{1}{4}\,g^{\alpha\beta}\,F^{\lambda\mu}\,F_{\lambda\mu}\right),
\end{aligned}
\tag{5.147}
$$

or

$$
\boxed{\Phi^\alpha \equiv -\partial_\beta\,T_{\text{em}}^{\alpha\beta}}\,,
\tag{5.148}
$$

where

$$
\boxed{4\pi\,T_{\text{em}}^{\alpha\beta} \equiv F^\alpha{}_\mu\,F^{\mu\beta} + g^{\alpha\beta}\,F^{\lambda\mu}\,F_{\lambda\mu}/4}
\tag{5.149}
$$

defines the *electromagnetic energy-momentum tensor* $T_{\text{em}}^{\alpha\beta}$. As was the case for its nonrelativistic counterpart (Equation [3.55]), the projection of this tensor normal to a surface, i.e., $\vec{T}_{\text{em}}^i/4\pi$, is the pressure in the direction i (see below).

These equations are elegant, but what do they mean? Other than in regions where the field couples to matter via an exchange of energy and momentum, Equation (5.148) reduces to a *continuity equation* that describes the conservation of field momentum (through the spatial component) and field energy (through the time component) at every point of spacetime. In other words, the expression

$$
\partial_\beta\,T_{\text{em}}^{\alpha\beta} = \frac{1}{c}\,F^\alpha{}_\mu\,J^\mu
\tag{5.150}
$$

reduces to a continuity equation for the field

$$
\partial_\beta\,T_{\text{em}}^{\alpha\beta} = 0\,,
\tag{5.151}
$$

analogous to the charge current equation $\partial_\alpha\,J^\alpha = 0$, when $J^\mu = 0$.

It is not difficult to show that

$$
T_{\text{em}}^{00} = \frac{1}{8\pi}\left(\mathbf{E}^2 + \mathbf{B}^2\right),
\tag{5.152}
$$

and that

$$
T_{\text{em}}^{0i} = \frac{1}{4\pi}(\mathbf{E}\times\mathbf{B})_i\,.
\tag{5.153}
$$

Thus, the 0th term of Equation (5.151) is really nothing more than

$$
\frac{1}{c}\,\frac{\partial T^{00}}{\partial t} + \frac{\partial T^{0i}}{\partial x^i} = 0\,,
\tag{5.154}
$$

or
$$\frac{\partial u}{\partial t} + \vec{\nabla} \cdot \mathbf{S} = 0 \,, \tag{5.155}$$

where
$$u \equiv \frac{1}{8\,\pi}(\mathbf{E}^2 + \mathbf{B}^2) \,, \tag{5.156}$$

and
$$\mathbf{S} \equiv \frac{c}{4\,\pi}\,(\mathbf{E} \times \mathbf{B}) \,, \tag{5.157}$$

exactly the conservation of field energy that we derived within the nonrelativistic context. In the same vein, the spatial part of Equation (5.151) results in the expression
$$\partial_0\,T_{\mathrm{em}}^{i0} + \partial_j\,T_{\mathrm{em}}^{ij} = 0 \,, \tag{5.158}$$

or
$$\frac{\partial g^i}{\partial t} - \vec{\nabla} \cdot \left(-\vec{T}_{\mathrm{em}}^{i}\right) = 0 \,, \tag{5.159}$$

where
$$\mathbf{g} \equiv \frac{1}{4\,\pi\,c}\,(\mathbf{E} \times \mathbf{B}) \tag{5.160}$$

is the electromagnetic momentum density and $\vec{T}_{\mathrm{em}}^{i}$ is the *force per unit area* (i.e., the *pressure*) in the direction i. (These expressions employ the *dyadic* notation for $T^{\alpha\beta}$ that we introduced in § 3.3.3.) So Equation (5.159) is the statement that there exists a momentum density of the field that satisfies a Newtonian equation of motion, i.e., the concept of electromagnetic momentum density is established from the field equations. Integrating $\vec{T}_{\mathrm{em}}^{i}$ over a surface S gives the component of the electromagnetic force acting in the direction i:
$$f^i = \int_S \vec{T}_{\mathrm{em}}^{i} \cdot d\mathbf{a} \,. \tag{5.161}$$

Note that this is fully consistent with our earlier definition of f^{α} in Equation (5.139), since the fields used here are those measured relative to the surface S, for which $\gamma = 1$. If there are no enclosed charges, then there must be as much electric flux entering the volume as there is leaving, which gives a net value of zero for the integral on the right-hand side of this equation, and therefore $f^i = 0$, as expected. This idea is central to our understanding of how $\vec{T}_{\mathrm{em}}^{i}$ gives the force. The total force $\mathbf{F}$ depends on the net flux represented by $\vec{T}_{\mathrm{em}}^{i}$ threading the enclosed area. Changes in $\vec{T}_{\mathrm{em}}^{i}$ result from the presence of sources, so the absence of enclosed charges implies the absence of a force.

Example 5.1. As an application of these concepts, we will revisit the problem we considered in § 2.1.1, where a charge q lies near an infinite plane conductor maintained at zero potential (Figure 5.5). We learned earlier that this problem is most easily and directly solved by using guesses and the symmetry of the situation to infer that the effects of the plane boundary could be simulated by invoking an imaginary charge $-q$ on the outside of the volume of solution.

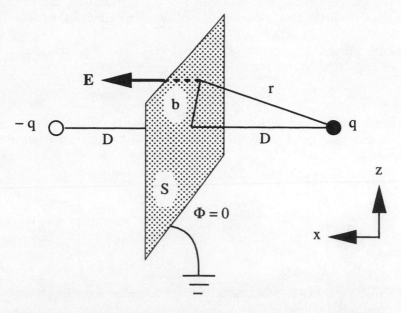

Figure 5.5 Geometry for calculating the force on a conducting boundary in the presence of a charge q, using the electromagnetic energy-momentum tensor, which represents the force per unit area on the surface enclosing the volume of interest. In this problem, the relevant volume is the space to the right of the boundary.

In Figure 5.5, this region of validity corresponds to the half-space $x < 0$. Although we didn't carry our previous solution beyond the step of calculating the potential, it should be obvious that determining the force on the boundary simply requires that we calculate the equivalent force on the charge $-q$, which by Coulomb's law has a magnitude

$$F_{-q} = \frac{q^2}{(2D)^2} \, . \tag{5.162}$$

So why are we still interested in this problem? The point is that here is an example of a situation where we know and understand how to calculate the force on a bounded region due to the presence of a charge which produces a field that threads the bounding surface. In other words, we know the answer and we want to apply our new formalism involving the energy-momentum tensor to see if we can reproduce it using a new physical principle. That is, we want to see if calculating the total *stress* on a surface S due to a flux of $\mathbf{E}$ and $\mathbf{B}$ through it gives the same force on the enclosed volume as if we simply used Coulomb's law on the charges within. We recall that this is one of the key characteristics that endows the fields with a dynamical reality, independent of the sources.

Because the plane is a conductor, $\mathbf{E}$ is perpendicular to the surface at $x = 0$. Thus, we know that on this plane,

$$\mathbf{E} = 2\frac{D}{r}\frac{q}{r^2}\,\hat{x} \equiv E^x\,\hat{x}\,, \qquad (5.163)$$

where the factor in the front on the right-hand side includes the two charges (real and imaginary) and the cosine of the projection angle perpendicular to the surface. Thus,

$$T_{\text{em}}^{\alpha\beta} = \frac{1}{4\pi}\begin{pmatrix} \frac{1}{2}(E^x)^2 & 0 & 0 & 0 \\ 0 & \frac{1}{2}(E^x)^2 & 0 & 0 \\ 0 & 0 & -\frac{1}{2}(E^x)^2 & 0 \\ 0 & 0 & 0 & -\frac{1}{2}(E^x)^2 \end{pmatrix}. \qquad (5.164)$$

In this situation, the elemental area is $d\mathbf{a} = da\,\hat{x}$, and so according to Equation (5.161), the force in direction i is

$$f^i = \int_S T_{\text{em}}^{ix}\,da\,. \qquad (5.165)$$

We see right away that $f^y = f^z = 0$. In addition,

$$\begin{aligned} f^x &= \int_S T_{\text{em}}^{xx}\,da \\ &= \frac{1}{8\pi}\int_S (E^x)^2\,da \\ &= \frac{q^2 D^2}{2\pi}\int_S \frac{da}{r^6}\,, \end{aligned} \qquad (5.166)$$

and since $da = 2\pi b\,db = \pi\,db^2$, and $r^6 = (D^2 + b^2)^3$, we get finally

$$f^x = \frac{q^2 D^2}{2}\int_0^\infty \frac{du}{(D^2 + u)^3} = \frac{q^2}{(2D)^2}\,, \qquad (5.167)$$

which is identical to F_{-q} as given in Equation (5.162).

6

THE LAGRANGIAN FORMULATION OF ELECTRODYNAMICS

6.1 ACTION PRINCIPLES IN CLASSICAL FIELD THEORIES

Our reformulation of electrodynamics in terms of 4D quantities in four-dimensional spacetime was motivated by a desire to produce a self-consistent description of this theory with the "new" (special relativistic) mechanics. The application of Galilean transformations to the Maxwell equations clearly produces relationships between the field components that are wrong in the case of relativistic systems with $v \to c$. In contrast, our interest in a Lagrangian formulation of Maxwell's equations, which by the way *builds upon the special relativistic treatment,* has nothing to do with a need to remove deficiencies in the theory. As we shall see, the sole motivation for using action principles is to "improve" our understanding of the underlying physics, with a goal of extracting additional laws that might not otherwise be apparent.

As an example of how a different perspective has served this purpose before, consider that the Newtonian formulation of classical mechanics is a description of the particle dynamics in terms of *forces*. But now look at the advantage of also introducing the concept of a potential energy

$$U = q\,\Phi\,, \tag{6.1}$$

from which the forces are derivable, i.e.,

$$\mathbf{F} = -\vec{\nabla} U\,. \tag{6.2}$$

For one thing, a description in terms of Φ allows us to define a *conserved* energy.

145

To see how this works in special relativity (see Chapter 5), let us consider the equation of motion for a charge q subjected to external electric and magnetic fields, viz.,

$$\frac{d\mathbf{p}}{dt} = \frac{d}{dt}(\gamma\, m\, \mathbf{v}) = q\left(\mathbf{E} + \vec{\beta} \times \mathbf{B}\right),$$ (6.3)

where $\vec{\beta} \equiv \mathbf{v}/c$. Evidently,

$$\mathbf{v} \cdot \frac{d\mathbf{p}}{dt} = q\,\mathbf{v} \cdot \mathbf{E}\,.$$ (6.4)

But

$$
\begin{aligned}
\mathbf{v} \cdot \frac{d\mathbf{p}}{dt} &= \gamma m\mathbf{v} \cdot \frac{d\mathbf{v}}{dt} + m\gamma^3 \frac{\mathbf{v}}{c^2} \cdot \frac{d\mathbf{v}}{dt}\,|\mathbf{v}|^2 \\[2mm]
&= \gamma m\mathbf{v} \cdot \frac{d\mathbf{v}}{dt}\left(1 + \frac{|\mathbf{v}|^2}{c^2}\gamma^2\right) \\[2mm]
&= \gamma^3 m\,\mathbf{v} \cdot \frac{d\mathbf{v}}{dt} \\[2mm]
&= m\,c^2\,\gamma^3\,\frac{\mathbf{v}}{c^2} \cdot \frac{d\mathbf{v}}{dt} \\[2mm]
&= \frac{d}{dt}\left(\gamma m c^2\right),
\end{aligned}
$$ (6.5)

which means that

$$\frac{d}{dt}(\gamma\, m\, c^2) = q\,\mathbf{v} \cdot \mathbf{E}\,.$$ (6.6)

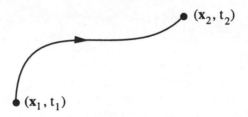

Figure 6.1   Particle trajectory from spacetime point $(\mathbf{x}_1, t_1)$ to $(\mathbf{x}_2, t_2)$.

We can integrate this expression over time along the particle's trajectory from t_1 to t_2 (Figure 6.1), and if the electric field is static, we then get

$$[\gamma\, m\, c^2]_2 - [\gamma\, m\, c^2]_1 = -q\,\Phi_2 + q\,\Phi_1\,.$$ (6.7)

That is, the quantity

$$\boxed{\mathcal{E} \equiv \gamma\, m\, c^2 + q\, \Phi} \tag{6.8}$$

is constant during the particle's motion as long as the electromagnetic field is static.

Yet another perspective on (or reformulation of) mechanics is provided by *Hamilton's variational principle*,

$$\delta I = 0 \,, \tag{6.9}$$

where

$$I \equiv \int_{t_1}^{t_2} L \, dt \,. \tag{6.10}$$

The integral I is thought of as some kind of generalized *action* (or perhaps it may even be thought of as an *expenditure* of "something") during the system's motion. The expenditure *rate*, L, which is known as the Lagrangian, characterizes the system and the circumstances in which the motions are to take place. The symbol δI stands for the variation of I from the value it takes during the actual motions that will occur. The expression $\delta I = 0$ is a condition for the actual action to be a comparative extremum. The variables used to construct L may be any quantities that might be observed to vary during the motion: $q_1(t)$, $q_2(t), \ldots, q_f(t)$, where f is the number of degrees of freedom of the system. For example, $q_i(t)$ could be a component of the position vector $\mathbf{x}_j(t)$ of particle j, or it could be an angle $\theta_i(t)$.

Classical mechanics can predict an explicit, unique motion only if the positions and velocities are known. Thus, the motion is specified by a set of $2f$ values: $q(t) \equiv q_1(t) \ldots q_f(t)$ and $\dot{q}(t) \equiv \dot{q}_1(t) \ldots \dot{q}_f(t)$. As such, the Lagrangian (i.e., the expenditure rate) at a given time t [or phase $q_i(t)$, $\dot{q}_i(t)$] is

$$L = L(q, \dot{q}, t) \,. \tag{6.11}$$

The system progresses through a sequence of phases (i.e., a *path*), and associated with this path will be a generalized action, as we have defined it (Figure 6.2).

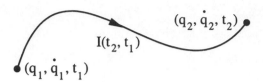

Figure 6.2 Particle trajectory from the generalized coordinate point $(q_1, \dot{q}_1, t_1)$ to $(q_2, \dot{q}_2, t_2)$. Each such path is associated with a generalized action $I(t_2, t_1)$ that measures the "expenditure" of the trajectory.

By forcing I to be an extremum, we are *forcing* a relationship between the $q(t)$ and the $\dot{q}(t)$ such that variations

$$q(t) \rightarrow q(t) + \delta q(t) ,\qquad (6.12)$$

and

$$\dot{q}(t) \rightarrow \dot{q}(t) + \delta \dot{q}(t) ,\qquad (6.13)$$

result in a $\delta I = 0$. *We are thereby calculating the actual equation of motion of the system!* To see how this works in practice, consider

$$\delta I = \int_{t_1}^{t_2} dt\, \delta L ,\qquad (6.14)$$

where

$$\begin{aligned}
\delta L &= \sum_i \left\{ \frac{\partial L}{\partial q_i} \delta q_i + \frac{\partial L}{\partial \dot{q}_i} \frac{d}{dt}(\delta q_i) \right\} \\
&= \frac{d}{dt}\left(\sum_i \frac{\partial L}{\partial \dot{q}_i} \delta q_i \right) + \sum_i \delta q_i \left(\frac{\partial L}{\partial q_i} - \frac{d}{dt}\frac{\partial L}{\partial \dot{q}_i} \right) .
\end{aligned}\qquad (6.15)$$

Thus,

$$\delta I = \left\{ \sum_i \frac{\partial L}{\partial \dot{q}_i} \delta q_i \right\}_{t_1}^{t_2} + \int_{t_1}^{t_2} dt \sum_i \delta q_i \left(\frac{\partial L}{\partial q_i} - \frac{d}{dt}\frac{\partial L}{\partial \dot{q}_i} \right) .\qquad (6.16)$$

But any motion between definite configurations $q(t_1)$ and $q(t_2)$ has $\delta q(t_1) = \delta q(t_2) = 0$, so that

$$\delta I = \int_{t_1}^{t_2} dt \sum_i \delta q_i \left(\frac{\partial L}{\partial q_i} - \frac{d}{dt}\frac{\partial L}{\partial \dot{q}_i} \right) = 0 .\qquad (6.17)$$

This can be satisfied at all times $t_1 < t < t_2$ and for any arbitrary variations δq_i in each degree of freedom only if

$$\boxed{\frac{d}{dt}\left(\frac{\partial L}{\partial \dot{q}_i}\right) - \frac{\partial L}{\partial q_i} = 0}\;, \tag{6.18}$$

for each $i = 1, 2, \ldots, f$. These are the well-known *Lagrange equations of motion*.[19]

Finding a correspondence between this formalism and the Newtonian equations

$$m \frac{d^2 \mathbf{x}_i}{dt^2} = \mathbf{F}_i \tag{6.19}$$

is particularly simple when the forces are derivable from a potential energy $U(\mathbf{x}_1, \mathbf{x}_2, \ldots, \mathbf{x}_f)$. Then, the *choice* of expenditure rate (i.e., Lagrangian)

$$L = T - U = \sum_i \frac{1}{2} m_i \left[\left(\frac{dx_i}{dt}\right)^2 + \left(\frac{dy_i}{dt}\right)^2 + \left(\frac{dz_i}{dt}\right)^2\right] - U \tag{6.20}$$

leads to

$$\frac{\partial L}{\partial \dot{x}_i} = m_i \dot{x}_i\;, \tag{6.21}$$

so that

$$m_i \frac{d^2 x_i}{dt^2} = -\frac{\partial U}{\partial x_i} = (\mathbf{F}_i)_x\;. \tag{6.22}$$

The quantity

$$\boxed{p_i \equiv \partial L / \partial \dot{q}_i} \tag{6.23}$$

is known as the *generalized* momentum, *conjugate* to the degree of freedom q_i. For example, when q_i is an angular degree of freedom, p_i becomes an angular momentum.

---

[19]For a more detailed treatment of this discussion, see Goldstein (1980).

6.2   RELATIVISTIC LAGRANGIANS OF POINT-CHARGE MOTIONS

A Lagrangian that makes the Lagrange equations equivalent to the equations of motion is simple to construct after a potential description of the field has been introduced. At this point, we should emphasize that although $\mathbf{B}$ is not an energy-changing agent, it is nonetheless a *momentum-changing* one (because it can change the direction of $\mathbf{v}$) and must therefore be included in the description.

We begin with the equation of motion (6.3), and write

$$\mathbf{E} = -\vec{\nabla}\Phi - \frac{1}{c}\frac{\partial \mathbf{A}}{\partial t} \,, \tag{6.24}$$

as before. For the second term, we can put

$$\begin{aligned} q\frac{\mathbf{v}}{c} \times \mathbf{B} &= \frac{q}{c}\mathbf{v} \times (\vec{\nabla} \times \mathbf{A}) \\ &= \frac{q}{c}\left\{ \vec{\nabla}(\mathbf{v} \cdot \mathbf{A}) - (\mathbf{v} \cdot \vec{\nabla})\mathbf{A} \right\}. \end{aligned} \tag{6.25}$$

But by the chain rule,

$$-(\mathbf{v} \cdot \vec{\nabla})\mathbf{A} = -\frac{d\mathbf{A}}{dt} + \frac{\partial \mathbf{A}}{\partial t} \,, \tag{6.26}$$

and so, combining Equations (6.3), (6.24), and (6.26), we arrive at the very important expression

$$\boxed{ \frac{d}{dt}\left(\gamma m \mathbf{v} + \frac{q}{c}\mathbf{A} \right) = -q\vec{\nabla}\left(\Phi - \frac{\mathbf{v} \cdot \mathbf{A}}{c} \right) } \,. \tag{6.27}$$

The Lagrangian $L(\mathbf{x}, \mathbf{v}, t)$ must be so constructed that each of the rectangular components of (6.27) takes on the form of Equation (6.18). But how do we do this? To begin with, the form of (6.27) is such that it is natural to associate the quantity

$$\mathbf{p}(\mathbf{x}, t) \equiv \gamma m \mathbf{v} + \frac{q}{c}\mathbf{A} \tag{6.28}$$

with the "generalized momentum," since it is the time derivative of this that changes in response to the gradient of a "potential" $q\left(\Phi - \mathbf{v} \cdot \mathbf{A}/c\right)$, i.e., the "generalized force." It is therefore reasonable to expect that

$$\frac{\partial L}{\partial v^i} = \gamma\, m\, v^i + \frac{q}{c}\, A^i \,, \tag{6.29}$$

and

$$\frac{\partial L}{\partial x^i} = -q\, \frac{\partial}{\partial x^i}\, \left(\Phi - \frac{\mathbf{v} \cdot \mathbf{A}}{c}\right) \,, \tag{6.30}$$

which will indeed satisfy the Lagrange equations. After some trial and error (as is the common procedure for finding a Lagrangian), we identify the following expression for L that is consistent with both (6.29) and (6.30):

$$\boxed{L = -\frac{m\, c^2}{\gamma} - q\, \left\{\Phi - \frac{\mathbf{v} \cdot \mathbf{A}}{c}\right\}} \,. \tag{6.31}$$

This is a *relativistic Lagrangian* in the sense that it makes the Lagrange equations reduce to the relativistic equations of motion.

We remark that the conjugate momentum of a charge in the presence of an electromagnetic field includes the term $q\,\mathbf{A}/c$—it is not just $\gamma\, m\, \mathbf{v}$! This additional term represents a field momentum available to q, just as $q\Phi$ serves as a storage of field energy available for exchange with the particulate kinetic energy. Both the kinetic momentum $\gamma\, m\, \mathbf{v}$ and the amount of $q\,\mathbf{A}/c$ stored as field momentum may be exchanged whenever the particle encounters gradients in the interaction energy $U(\mathbf{x}) = q\,(\Phi - \mathbf{v} \cdot \mathbf{A}/c)$ of the particle and field.

Note that the Lagrangian in Equation (6.31) depends explicitly on the potentials, so it is not invariant under a gauge transformation, such as that given in Equations (3.10) and (3.11). It is trivial to see from (6.31) that L acquires a term $d\psi/d(ct)$ under this transformation. However, total time derivatives in L do not alter the action integral, nor the Euler-Lagrange equations derived from it (see, e.g., Goldstein 1980). Thus, although this Lagrangian is not invariant under a gauge transformation, it nonetheless yields the correct (gauge invariant) equations of motion. In the same vein, the term $q\,\mathbf{A}/c$ appearing in the conjugate momentum is itself not gauge invariant, so it should not contribute to the particle's energy. In fact, using the standard definition of a Hamiltonian, i.e., $H \equiv \mathbf{p} \cdot \mathbf{v} - L$, one sees right away (using Equation [6.28] to evaluate $\mathbf{v}$) that H represents the correct special relativistic energy of a particle with potential energy $q\Phi$.

It is also of interest to note that a neutral point particle, or a charged particle in a field-free region of spacetime, is associated with the neutral-particle Lagrangian

$$L_0(\mathbf{v}) = -\frac{m\,c^2}{\gamma}\,. \tag{6.32}$$

Thus, Hamilton's variational principle for this simplified situation reduces to

$$\delta \int_{t_1}^{t_2} \frac{d\,t}{\gamma(\mathbf{v})} = \delta \int_{\tau(t_1)}^{\tau(t_2)} d\tau = 0\,. \tag{6.33}$$

That is, it becomes a principle of *least proper time* for the motion in free space!

This formalism also has a natural correspondence with the description of mechanics in terms of energy. Consider the derivative

$$\begin{aligned}
\frac{d}{d\,t} L(q, \dot{q}, t) &= \sum_i \left(\frac{\partial L}{\partial q_i} \dot{q}_i + \frac{\partial L}{\partial \dot{q}_i} \frac{d^2 q_i}{d\,t^2} \right) + \frac{\partial L}{\partial t} \\
&= \sum_i \left(\frac{d p_i}{d\,t} \frac{d q_i}{d\,t} + p_i \frac{d^2 q_i}{d\,t^2} \right) + \frac{\partial L}{\partial t}\,,
\end{aligned} \tag{6.34}$$

where we have used the Lagrange equations of motion to write

$$\frac{\partial L}{\partial q_i} = \frac{d}{d\,t} \left(\frac{\partial L}{\partial \dot{q}_i} \right) = \frac{d p_i}{d\,t}\,. \tag{6.35}$$

Then,

$$\frac{d\,H}{d\,t} = -\frac{\partial L}{\partial t}\,, \tag{6.36}$$

where

$$H \equiv \sum_i p_i\,\dot{q}_i - L\,. \tag{6.37}$$

Clearly, H is a *conserved* quantity whenever L depends only on the coordinates q and $\dot{q}$ of the motion, and not explicitly on the time at which these coordinates are realized.

For the simple example

$$L = T - U = \sum_i \frac{1}{2} m_i \left(\dot{x}_i^2 + \dot{y}_i^2 + \dot{z}_i^2 \right) - U\,, \tag{6.38}$$

we get $\mathbf{p}_i = m_i\,d\mathbf{x}_i/d\,t = m_i\,\mathbf{v}_i$, so that

$$H = \sum_j \mathbf{p}_j \cdot \mathbf{v}_j - L = \sum_j \frac{1}{2} m_j\,v_j^2 + U\,, \tag{6.39}$$

the familiar nonrelativistic energy. We identify H with a *generalized energy* that is conserved when $\partial L/\partial t = 0$.

In the case of a relativistic particle interacting with the electromagnetic field,

$$
\begin{aligned}
H &\equiv \mathbf{p} \cdot \mathbf{v} - L \\
&= \left(\gamma \, m \, \mathbf{v} + \frac{q}{c} \, \mathbf{A} \right) \cdot \mathbf{v} + \frac{m \, c^2}{\gamma} + q \left(\Phi - \frac{\mathbf{v} \cdot \mathbf{A}}{c} \right) \\
&= \gamma \, m \, c^2 \left(\frac{|\mathbf{v}|^2}{c^2} + \frac{1}{\gamma^2} \right) + q \Phi \, .
\end{aligned}
\tag{6.40}
$$

That is,

$$
\boxed{H = \gamma \, m \, c^2 + q \Phi} \, .
\tag{6.41}
$$

This quantity is conserved if $\partial L/\partial t = -\partial H/\partial t = -q \partial \Phi/\partial t = 0$, i.e., in the presence of a static field. (Note that an electromagnetic field cannot support $\partial \Phi/\partial t = (\mathbf{v}/c) \cdot \partial \mathbf{A}/\partial t$ for all values of $\mathbf{v}$.) After all, the energy is not expected to be conserved if Φ varies while the particle is moving.

6.3   THE FIELD LAGRANGIAN

Thus far in this chapter, we have considered the Lagrange formulation of the equations of motion for a charged particle in the presence of an external electromagnetic field. Here, we examine the corresponding Lagrangian description of the electromagnetic field itself interacting with external sources of charge and electric current. Before doing so, however, it is worth noting that the action I must be a *Lorentz scalar* because the equations of motion are derivable from the extremum condition $\delta I = 0$. By postulate (1) of special relativity, I cannot be a function of velocity. Thus, since

$$
I \equiv \int_{t_1}^{t_2} L \, dt = \int d^3 x \int_{t_1}^{t_2} dt \, \mathcal{L} \, ,
\tag{6.42}
$$

where $\mathcal{L}$ is the Lagrangian density, we can write

$$
I = \int d^4 x \, \mathcal{L} \, .
\tag{6.43}
$$

(Note that this differs from the earlier definition of I by a constant factor c, which will not effect the variational analysis.) But the volume element d^4x is also invariant (i.e., $[dx\,dy\,dz/\gamma][\gamma\,dt] = \gamma\,dx\,dy\,dz\,d\tau$), so $\mathcal{L}$ must itself be a Lorentz scalar.

In applying the Lagrangian approach to continuous fields, the generalized coordinate q_i is replaced by a continuous field $\phi^k(x)$, where instead of one discrete index i, we now have a discrete index k, and a continuous index x^α designating spacetime coordinates. The generalized velocity $\dot{q}_i$ is replaced by the 4-vector gradient $\partial_\beta\,\phi^k(x)$. For example, in the case of an electromagnetic field,

$$\phi^k(x) = A^k(x) \qquad (k = 0, 1, 2, 3) \,, \tag{6.44}$$

where $A^\alpha \equiv (\Phi, \mathbf{A})$. Thus, the Lagrangian density may be written

$$\mathcal{L} = \mathcal{L}\,[A^\alpha(x), \partial_\alpha\,A^\beta(x), x] \,, \tag{6.45}$$

in correspondence with the particle Lagrangian density $\mathcal{L}\,[q_i(t), \dot{q}_i(t), t]$. Variation of the action now results in variation of the variables

$$A^\alpha(x) \to A^\alpha(x) + \delta A^\alpha(x) \,, \tag{6.46}$$

and

$$\delta(\partial_\alpha\,A^\beta) = \partial_\alpha[\delta\,A^\beta(x)] \tag{6.47}$$

at each spacetime point within the 4-volume $\int d^4x$. Thus,

$$\begin{aligned}
\delta\mathcal{L} &= \frac{\partial\mathcal{L}}{\partial A^\nu}\,\delta A^\nu + \frac{\partial\mathcal{L}}{\partial(\partial_\mu\,A^\nu)}\,\partial_\mu(\delta A^\nu) \\[2ex]
&= \partial_\mu\left\{\frac{\partial\mathcal{L}}{\partial(\partial_\mu\,A^\nu)}\,\delta A^\nu\right\} \\[2ex]
&\quad + \delta A^\nu\left\{\frac{\partial\mathcal{L}}{\partial A^\nu} - \partial_\mu\frac{\partial\mathcal{L}}{\partial(\partial_\mu\,A^\nu)}\right\} \,.
\end{aligned} \tag{6.48}$$

As always, a repeated index denotes a summation over its value, so for example, $\partial_\mu\,A^\mu \equiv \sum_\mu \partial_\mu\,A^\mu$. As before, integration of $\delta\mathcal{L}$ over d^4x results in the evaluation of some $\delta\,A^\alpha(x)$ on the boundary of $\int d^4x$. But $\delta A^\alpha(x) = 0$ there to maintain consistency with the boundary conditions, and so the variation $\delta\,I$ will be zero for arbitrary variations within the spacetime volume only if

$$\boxed{\partial_\mu\frac{\partial\mathcal{L}}{\partial(\partial_\mu\,A^\nu)} = \frac{\partial\mathcal{L}}{\partial A^\nu}} \,, \tag{6.49}$$

for $\nu = 0, 1, 2, 3$. These now become the Euler-Lagrange equations for the electromagnetic field.

But we know that in order for $A^\alpha(x)$ to be the potential field arising from a given source distribution $J^\alpha(x)$, the Lagrange equations must be equivalent to the wave equation for $A^\alpha(x)$, which we derived earlier using the classical formalism. In the *Lorenz gauge*, defined by the condition

$$\partial_\alpha A^\alpha = 0 \,, \tag{6.50}$$

this equation says that

$$\partial^\mu \partial_\mu A^\alpha(x) = \frac{4\pi}{c} J^\alpha(x) \,. \tag{6.51}$$

Thus, the quantities A^α and $\partial_\alpha A^\beta$ must occur in such a way that

$$\frac{\partial \mathcal{L}}{\partial(\partial_\alpha A^\beta)} = \frac{1}{4\pi} \partial^\alpha A_\beta(x) \,, \tag{6.52}$$

and

$$\frac{\partial \mathcal{L}}{\partial A^\alpha} = \frac{1}{c} J_\alpha \,. \tag{6.53}$$

One form of $\mathcal{L}$ that satisfies both of these constraints is

$$\mathcal{L} = \frac{1}{16\pi}(\partial_\lambda A_\nu - \partial_\nu A_\lambda)(\partial^\lambda A^\nu - \partial^\nu A^\lambda) + \frac{1}{c} J_\alpha A^\alpha \,, \tag{6.54}$$

or equivalently,

$$\boxed{\mathcal{L} = F_{\alpha\beta} F^{\alpha\beta}/16\pi + J_\alpha A^\alpha/c} \,. \tag{6.55}$$

It is trivial to show that the wave equation (6.51) is equivalent to the inhomogeneous Maxwell equations and that the latter therefore follow naturally from the Lagrange equations with the Lagrangian density specified in (6.55). In the Lorenz gauge, we may write

$$\partial^\mu(\partial_\mu A^\alpha) = \partial_\mu(\partial^\mu A^\alpha) - \partial^\alpha(\partial_\mu A^\mu) \,. \tag{6.56}$$

Thus,

$$\begin{aligned}
\partial^\mu(\partial_\mu A^\alpha) &= \partial_\mu \{\partial^\mu A^\alpha - \partial^\alpha A^\mu\} \\
&= \partial_\mu F^{\mu\alpha} \\
&= \frac{4\pi}{c} J^\alpha \,.
\end{aligned} \tag{6.57}$$

The homogeneous equations are satisfied automatically because of the definition of $F^{\alpha\beta}$.

To close this subsection, we remark that the noninteracting (or "free-field") Lagrangian density results from the condition $J^\alpha = 0$ and is

$$\mathcal{L}_0 = \frac{1}{16\,\pi}\, F_{\alpha\beta}\, F^{\alpha\beta} \ . \tag{6.58}$$

But

$$F_{\alpha\beta}\, F^{\alpha\beta} = 2(\mathbf{B}^2 - \mathbf{E}^2) \ , \tag{6.59}$$

whence

$$\mathcal{L}_0 = \frac{\mathbf{B}^2 - \mathbf{E}^2}{8\,\pi} \ . \tag{6.60}$$

Thus, just as a particle's motion results from an exchange of its kinetic and potential energy, redistributions of a field isolated from external sources are due to exchanges between the electric and magnetic field energy distributions. The term $J_\alpha\, A^\alpha/c = \rho\Phi - \mathbf{J}\cdot\mathbf{A}/c$ is therefore appropriately a term that arises from the *interaction* between the particles and fields.

6.4   INVARIANCES AND CONSERVATION LAWS (NOETHER'S THEOREM)

It is in this subsection that we may start to realize the goal we expressed at the beginning of this chapter, i.e., to build upon the special relativistic formulation of electrodynamics with the Hamiltonian variational principle in an attempt to extract additional physical laws that might not otherwise be apparent. Not surprisingly (by analogy with classical mechanics), we will discover that a symmetry, or an invariance, of the action I finds expression in the form of a conservation law.

A natural expectation of the first postulate of special relativity is that the phase space trajectory of a system is independent of the starting time and of the absolute positioning in space, as long as the boundary conditions are kept fixed. That is, the system is expected to be left undisturbed by a shift of a constant 4-vector Δ_α in the spacetime coordinates:

$$x^\alpha \to x'^{\,\alpha} = x^\alpha + \Delta^\alpha \ . \tag{6.61}$$

There will be an invariance associated with this shift if the transformed potential fields $A'^{\,\nu}(x')$ satisfy the Euler-Lagrange equations of motion, just as the

$A^\nu(x)$ do (Equation [6.49]), and if the same boundary conditions still apply. Invariance of the action requires that

$$I = I' = \int_{V_4'} d^4x' \, \mathcal{L}\left[A'^\nu(x'), \partial_\mu' A'^\nu, x'^\mu\right] . \tag{6.62}$$

A translation will not distort volume elements, so

$$d^4x' = d^4x . \tag{6.63}$$

Moreover, if x' and x are to describe the same spacetime point, invariance demands that

$$A'^\nu(x') = A^\nu(x) , \tag{6.64}$$

and since $\partial_\mu' = \partial_\mu$ for variables differing only by a constant, this also means that

$$\partial_\mu' A'^\nu = \partial_\mu A^\nu . \tag{6.65}$$

Thus,

$$\begin{aligned} I' - I &= \int d^4x \left[\mathcal{L}\left\{A^\nu(x), \partial_\mu A^\nu, x^\alpha + \Delta^\alpha\right\} \right. \\ &\quad \left. - \mathcal{L}\left\{A^\nu(x), \partial_\mu A^\nu, x^\alpha\right\} \right] . \end{aligned} \tag{6.66}$$

For an infinitesimal shift, Δ^α, the integrand is just

$$\mathcal{I} \equiv \sum_\alpha \left(\frac{\partial \mathcal{L}}{\partial x^\alpha}\right)_{A^\nu} \Delta^\alpha , \tag{6.67}$$

where differentiation is only with respect to the explicit dependence of $\mathcal{L}$ on x. However, we still cannot integrate Equation (6.66) because of the restriction on A^ν in (6.67). We must therefore rewrite this derivative as follows:

$$\left(\frac{\partial \mathcal{L}}{\partial x^\alpha}\right)_{A^\nu} = \frac{\partial \mathcal{L}}{\partial x^\alpha} - \frac{\partial \mathcal{L}}{\partial A^\beta} \partial_\alpha A^\beta - \frac{\partial \mathcal{L}}{\partial(\partial_\mu A^\nu)} \partial_\alpha(\partial_\mu A^\nu) . \tag{6.68}$$

Since $A^\alpha(x)$ is supposed to satisfy the Euler-Lagrange equations of motion, we can write

$$\frac{\partial \mathcal{L}}{\partial A^\nu} = \partial_\mu \left[\frac{\partial \mathcal{L}}{\partial(\partial_\mu A^\nu)}\right] , \tag{6.69}$$

so that

$$\left(\frac{\partial \mathcal{L}}{\partial x^\alpha}\right)_{A^\nu} = \frac{\partial \mathcal{L}}{\partial x^\alpha} - \partial_\mu \left[\frac{\partial \mathcal{L}}{\partial(\partial_\mu A^\nu)} \partial_\alpha A^\nu\right] . \tag{6.70}$$

Thus,

$$I' - I = \Delta^\alpha \int_{V_4} d^4x \, \partial_\mu \left[\mathcal{L} \, \delta^\mu{}_\alpha - \frac{\partial \mathcal{L}}{\partial(\partial_\mu A^\nu)} \, \partial_\alpha A^\nu \right] . \tag{6.71}$$

But Δ^α and V_4 are arbitrary, and we see finally that translational invariance implies

$$\partial_\mu \mathcal{H}^\mu{}_\alpha = 0 , \tag{6.72}$$

where

$$\mathcal{H}^\mu{}_\alpha \equiv \frac{\partial \mathcal{L}}{\partial(\partial_\mu A^\nu)} \, \partial_\alpha A^\nu - \mathcal{L} \, \delta^\mu{}_\alpha . \tag{6.73}$$

As suggested by the naming of this quantity, $\mathcal{H}$ has the same attributes as the Hamiltonian $H = p_i \, \dot{q}_i - L$ we encountered earlier in our discussion of particle dynamics, and it is thus expected that the vanishing divergence exhibited in Equation (6.72) constitutes a continuity equation describing the conservation of the field energy and momentum, which we now prove in the case of a free-field. When $J^\alpha = 0$, the Lagrangian density may be written

$$\begin{aligned} \mathcal{L}_0 &= +\frac{1}{16\pi} F^{\alpha\beta} F_{\alpha\beta} \\ &= \frac{1}{16\pi} (\partial^\alpha A^\beta - \partial^\beta A^\alpha)(\partial_\alpha A_\beta - \partial_\beta A_\alpha) , \end{aligned} \tag{6.74}$$

which results in the field Hamiltonian (also known in the literature as the *canonical stress tensor*)

$$\mathcal{H}^\mu{}_\alpha = \frac{1}{4\pi} \left\{ F^{\mu\nu} \, \partial_\alpha A_\nu - \frac{\delta^\mu{}_\alpha}{4} F^{\lambda\beta} F_{\lambda\beta} \right\} . \tag{6.75}$$

But

$$\partial_\alpha A_\nu = \partial_\nu A_\alpha - F_{\alpha\nu} = \partial_\nu A_\alpha + F_{\nu\alpha} , \tag{6.76}$$

so that

$$\mathcal{H}^\mu{}_\alpha = \frac{1}{4\pi} \left(F^{\mu\nu} F_{\nu\alpha} - \frac{1}{4} \delta^\mu{}_\alpha F^{\lambda\beta} F_{\lambda\beta} \right) + \frac{1}{4\pi} F^{\mu\nu} \, \partial_\nu A_\alpha . \tag{6.77}$$

Evidently,

$$\mathcal{H}^\mu{}_\alpha \equiv T_{\text{em}}{}^\mu{}_\alpha + \frac{1}{4\pi} F^{\mu\nu} \, \partial_\nu A_\alpha , \tag{6.78}$$

where $T_{\text{em}}{}^\mu{}_\alpha$ is the symmetric stress-energy-momentum tensor for the electromagnetic field (Equation [5.149]), containing the field energy-momentum densities and their fluxes. At this point we should mention that the canonical stress tensor depends on the choice of gauge. One can see, for example, that $\mathcal{H}^0{}_i = (\mathbf{E} \times \mathbf{B})_i / 4\pi + \vec{\nabla} \cdot (A_i \, \mathbf{E}) / 4\pi$, which is the standard momentum density except for the addition of a divergence term. Upon integration over all space, however, these added terms produce surface integrals at infinity where the fields and potentials are zero, so they give no contribution. We have, therefore,

$$\partial_\mu \, \mathcal{H}^\mu{}_\nu = \partial_\mu \, T_{\text{em}}{}^\mu{}_\nu + \frac{1}{4\,\pi} \, \partial_\mu \, (F^{\mu\lambda} \, \partial_\lambda \, A_\nu) \, . \qquad (6.79)$$

The last term can be written as

$$\partial_\mu (F^{\mu\lambda} \, \partial_\lambda \, A_\nu) = \partial_\mu \, \partial_\lambda \, (F^{\mu\lambda} \, A_\nu) - \partial_\mu (A_\nu \, \partial_\lambda \, F^{\mu\lambda}) \, , \qquad (6.80)$$

in which the second derivative on the right-hand side vanishes because of Maxwell's equation for a sourceless field, and the first derivative also vanishes because $F^{\mu\lambda}$ is antisymmetric. We have thus arrived at the expected result,

$$\boxed{\partial_\mu \, \mathcal{H}^\mu{}_\nu = \partial_\mu \, T_{\text{em}}{}^\mu{}_\nu = 0} \, , \qquad (6.81)$$

which is the continuity equation for a sourceless field. *Translational invariance implies, and is implied by, the conservation of field energy-momentum.*

In the same spirit, a system with an isolated electromagnetic field ought to be invariant to spatial rotations as long as it is screened from external influences. The physical implications of this invariance may be derived in a fashion similar to that for the translational invariance, and we shall omit the details for the sake of brevity. It is not difficult to show (and to understand!) that *the rotational invariance of the action implies, and is implied by, the conservation of field angular momentum.*

Finally, we shall use these tools to understand the physical basis for the *gauge invariance* of the electromagnetic field. We have suggested on at least two occasions (i.e., for the Coulomb gauge and the Lorenz gauge) that physically meaningful results should be invariant under gauge transformations of the potentials, expressed as

$$\bar{A}_\mu = A_\mu + \partial_\mu \chi \, . \qquad (6.82)$$

For example, as we saw in § 3.2, a set of potentials in the Lorenz gauge (for which $\partial_\mu A^\mu = 0$) will remain in this gauge under a transformation (6.82) as long as χ is a 4-scalar function satisfying the wave equation

$$\partial^\mu \, \partial_\mu \, \chi = 0 \, . \qquad (6.83)$$

In that case,

$$\partial^\mu \, \partial_\mu \, \bar{A}_\alpha = \partial^\mu \, \partial_\mu \, A_\alpha = \frac{4\,\pi}{c} \, J_\alpha \, . \qquad (6.84)$$

But notice that the Lagrangian for an *interacting* field *need not be* invariant under such a gauge transformation, since

$$\mathcal{L} = \frac{1}{16\,\pi} F^{\alpha\beta} F_{\alpha\beta} + \frac{1}{c} J^\alpha A_\alpha$$

$$\rightarrow \bar{\mathcal{L}} = \frac{1}{16\,\pi} F^{\alpha\beta} F_{\alpha\beta} + \frac{1}{c} J^\alpha A_\alpha + \frac{1}{c} J^\alpha \partial_\alpha \chi \, . \tag{6.85}$$

That is,

$$\bar{\mathcal{L}} - \mathcal{L} = \frac{1}{c} J^\alpha \partial_\alpha \chi \, . \tag{6.86}$$

A more useful form of this equation is

$$\bar{\mathcal{L}} - \mathcal{L} = \frac{1}{c} \partial_\alpha \left(J^\alpha \chi \right) - \frac{\chi}{c} \partial_\alpha J^\alpha \, , \tag{6.87}$$

because the action and equations of motion are not affected by the addition of a 4-divergence to the Lagrangian density, as may be verified by direct substitution of (6.87) into Equation (6.49). The only difference that remains between $\bar{\mathcal{L}}$ and $\mathcal{L}$ is a term proportional to $\partial_\alpha J^\alpha$! Thus, *these physical laws are invariant under this potential gauge transformation as long as $\partial_\alpha J^\alpha = 0$, i.e., as long as charge is absolutely conserved.* This is in fact true for any gauge, not just the Lorentz gauge that we have considered here for illustrative purposes. *Potential gauge invariance implies, and is implied by, the charge-conservation principle!*

7

RELATIVISTIC
TREATMENT OF
RADIATION

Our classical ideas regarding fields and their sources culminated with our attempts to understand the process of radiation in § 4.6. We came away with the feeling that we had understood at least the basics of how $\mathbf{E}$ and $\mathbf{B}$ behave once they "leave" the charges, and how they evolve from the so-called near fields into the far fields out in the radiation zone. But just as mechanics is altered significantly by special relativity, we might reasonably expect that the electromagnetic field could also be altered from its low-velocity form (Equations [4.166], [4.167]) when $\beta \to 1$. Retracing our steps in § 4.6.2 should convince us that no transformation from the particle to the lab frame (or vice versa) was ever invoked in our derivations. All quantities, such as ρ and $\mathbf{J}$, were those pertaining to the coordinate system in which the fields were being determined. So our calculation of the field should not be affected by the introduction of special relativity (since we are in the same frame as the sources), though its dynamical impact on the charges will of course be subject to the relativistic effects in the force equation (5.129). Toward the end of this chapter we will concern ourselves with the back-reaction of the radiation process on the sources, and we anticipate that relativistic effects will thus be important when the field and particle dynamics equations are solved self-consistently, since the particle trajectory is necessarily altered from its classical form due to the particle's velocity-dependent inertia. Special relativity also provides an explanation for why infinite values of $\mathbf{E}$ (which can occur in some directions when $\beta \to 1$) should be excluded. According to Equation (5.83), a particle must gain infinite energy for β to even approach 1, let alone surpass it, so it is not possible for any source to be accelerated to speeds larger than the speed of light. (However, this does not in itself argue against the possible existence of *tachions*, particles that always travel faster than light.)

In the next few sections, we shall consider several important issues, including (i) deriving an expression for the total radiation emitted, (ii) finding its angular distribution, and (iii) calculating its spectrum as a function of frequency. In our efforts to convert the mathematical description into the more complete and elegant language of four-dimensional spacetime, it may appear that we are retracing some of the steps we have already taken. This is definitely true in part, because the underlying physics is the same. For example, in our previous derivation, we divided out the time and space dependence of the wave equation by using a Fourier expansion for the potentials. In effect, we were solving first for what we would now call the dependence of Φ and $\mathbf{A}$ on the time component of x^α. The solution to the space piece of the equations subsequently led to the Liénard-Wiechert potentials (4.146) and (4.147). Since we now know that Φ and $\mathbf{A}$, like t and $\mathbf{x}$, are all components of the same physical entity in four-dimensional spacetime, this division (between Φ and $\mathbf{A}$ and between t and $\mathbf{x}$) should not be necessary. However, we will see that our earlier use of a Fourier expansion to handle the time dependence of the potentials is equivalent to our handling of k^0, which we treat as a complex variable in order to carry out the first part of the integration for the Green function. One of the benefits of rederiving the fields using the framework of special relativity is therefore to see how this natural union comes about. Having already discussed many of the principles entering into the treatment of radiation in § 4.6, we should be able to better understand the motivation for the ideas and techniques that we shall introduce here.

7.1 THE GREEN FUNCTION IN FOUR-DIMENSIONAL SPACETIME

Mirroring our classical derivation in § 3.2, we begin with the inhomogeneous Maxwell's equations, which describe the time-dependent electromagnetic field produced by a given configuration of sources. Ultimately, all the properties of the radiation field must be encompassed by the behavior of these equations, which we have now seen coalesce into a single covariant expression (5.119) in 4D spacetime. We remind ourselves that this is but one piece of the mathematical description of the interaction between the particles and fields—the other being the Lorentz force equation (5.129), which accounts for the effect on the charge due to the presence of the field—so that both must be considered simultaneously to produce a self-consistent description of the charge-field interaction. We shall return to this in §§ 7.5 and 7.6 below.

For now, we will consider the impact of the source equation, which in the Lorenz gauge $(\partial_\alpha A^\alpha = 0)$ may be written

$$\partial_\alpha \partial^\alpha A^\beta = \frac{4\pi}{c} J^\beta \tag{7.1}$$

(see Equation [6.57]). Although this expression may look very different from Equations (3.13) and (3.14), it contains very similar physics. What *is* different here is that space and time are now linked by the special relativistic constraint (5.44), resulting from the postulate that the speed of light is finite and invariant from frame to frame. That is, it's not so much that we must handle Φ and $\mathbf{A}$ (or more elegantly, A^α) differently, but rather it's that the differentiation variables $\mathbf{x}$ and t need to be treated in a covariant manner. The solution to Equation (7.1) must explicitly show a dependence on $\mathbf{x}$ and t that is consistent with the invariance of the spacetime interval Δs. This suggests that we again approach the problem using a time-dependent Green function technique, since we anticipate that the "retarded" nature of the classical solution will be preserved. By analogy with (3.18), we seek a function

$$G(x, x') , \qquad (x \equiv x^\alpha) , \tag{7.2}$$

such that

$$\frac{\partial}{\partial x^\alpha} \frac{\partial}{\partial x_\alpha} G(x, x') = 4\pi \delta^4(x - x') , \tag{7.3}$$

where

$$\delta^{(4)}(x - x') \equiv \delta(x^0 - x'^0) \, \delta^3(\mathbf{x} - \mathbf{x}') \tag{7.4}$$

is the 4D δ-function. In our classical treatment of this problem, we saw no evidence that the solution for Φ and $\mathbf{A}$ should depend on the relative angle between $\mathbf{x} - \mathbf{x}'$ and our arbitrary choice of coordinate frame, and we argued that G should therefore depend only on the magnitude of the vector $\mathbf{x} - \mathbf{x}'$, not its direction, i.e., $G(\mathbf{x}, \mathbf{x}') = G(\mathbf{x} - \mathbf{x}')$. Here, the situation is at first not as obvious, since we now know that length contractions and time dilations are not the same in the "boost" and perpendicular directions. Thus, it might appear that the 4D Green function could be angle dependent. But if we think about this carefully, we realize that the angle dependence would enter into the picture only during a transformation from frame to frame. Without specifying this transformation a priori, the field solution wouldn't "know" about it and so we make the guess that $G(x, x') = G(x - x')$, which we shall justify by showing that it yields a solution to Equation (7.1). So where *does* the angle dependence enter the problem? Clearly, it must affect the transformation of $\mathbf{x}$ and t, and it therefore determines the 4D orientation of the interval $x^\alpha - x'^\alpha$ in one frame relative to that in the other. Thus, we can write

$$\frac{\partial}{\partial z^\alpha} \frac{\partial}{\partial z_\alpha} G(z) = 4\pi \delta^4(z) , \tag{7.5}$$

where

$$z^\alpha \equiv x^\alpha - x'^\alpha . \tag{7.6}$$

To solve for $G(z)$, we can expand it in terms of its Fourier components $\tilde{G}(k)$, where

$$G(z) = \frac{1}{(2\pi)^4} \int d^4k \; \tilde{G}(k) \; \exp(-ik_\alpha z^\alpha) . \tag{7.7}$$

We must remember that we are now in 4D spacetime, so

$$k_\alpha z^\alpha = k_0 z^0 - \mathbf{k} \cdot \mathbf{z} . \tag{7.8}$$

Thus, in the wave equation,

$$
\begin{aligned}
\frac{\partial}{\partial z^\alpha} \frac{\partial}{\partial z_\alpha} G(z) &= \frac{1}{(2\pi)^4} \int d^4k (-ik_\mu)(-ik^\mu) \, \tilde{G}(k) \, \exp(-ik_\alpha z^\alpha) \\
&= -\frac{1}{(2\pi)^4} \int d^4k \, k^2 \tilde{G}(k) \exp(-ik_\alpha z^\alpha) \\
&= 4\pi \delta^4(z) = \frac{4\pi}{(2\pi)^4} \int d^4k \, \exp(-ik_\alpha z^\alpha) .
\end{aligned}
\tag{7.9}
$$

This last step follows from Equation (7.5) and the 4D Fourier transform of the Dirac delta function. Thus, we need

$$k_\alpha k^\alpha \, \tilde{G}(k) = -4\pi , \tag{7.10}$$

which leads to the elegant result

$$\boxed{\tilde{G}(k) = -4\pi/k^2} . \tag{7.11}$$

This form of $\tilde{G}(k)$ is known as the *propagator* (in momentum space) and is the predecessor of several others in physics pertaining to their respective fields. We therefore see that the Green function for this problem is

$$\boxed{G(x - x') = -\frac{4\pi}{(2\pi)^4} \int d^4k \, \frac{\exp\{-ik_\alpha(x^\alpha - x'^\alpha)\}}{k^2}} . \tag{7.12}$$

Unfortunately, this is still far from being a practical result because the integrand is singular:

$$G(x - x') = -\frac{4\pi}{(2\pi)^4} \int d^3k \, dk_0 \, \exp\{i\mathbf{k} \cdot (\mathbf{x} - \mathbf{x}')\} \frac{\exp\{-ik_0 \, (x^0 - x'^0)\}}{(k^0)^2 - \mathbf{k}^2} . \quad (7.13)$$

A straightforward technique for solving this integral employs contour integration in the complex plane, treating k^0 as the complex variable (Figure 7.1).

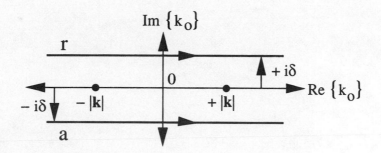

Figure 7.1 The complex k_0 plane. In order to evaluate the integral in Equation (7.13), we must incrementally lift the integration path off the real axis by an amount $\pm i\delta$ and then close the contour with a semicircle at infinity.

Lifting the integration segment along the real axis by an amount $\pm i\delta$ in the direction of $\text{Im}(k_0)$, we can then close the integration loop with a semicircle at infinity. However, we must be careful to choose a loop in either the positive or negative imaginary half-plane subject to the condition that the contribution to the integral from the semicircle is zero. Therefore, when $z^0 > 0$, we need $\exp(-ik_0 z^0) \to 0$ in the limit $|\text{Im}(k_0)| \to \infty$; i.e., when $z^0 > 0$, we need $\text{Im}(k_0) < 0$ so that the loop must be closed in the negative half-plane. When $z^0 < 0$, however, we must have $\text{Im}(k_0) > 0$ so that the loop in this instance should be closed in the positive half-plane. Clearly, depending on which of these contours we choose for either segment r or segment a will decide whether the loop encloses two singularities or none.

Before proceeding, let us take a short aside to review the residue theorem. If f(z) is analytic inside and on a closed curve C, except for isolated singular points $z_1, z_2, \ldots, z_N$ lying inside C, then

$$\oint_C f(z) \, dz = 2\pi i \sum_{k=1}^{N} \text{Res}_{z=z_k} f(z) , \quad (7.14)$$

where C is evaluated counterclockwise and $\text{Res}_{z=z_k} f(z)$ is the residue of $f(z)$ at z_k (Figure 7.2).

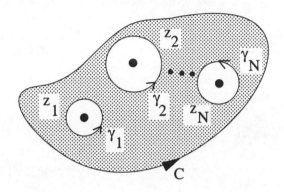

Figure 7.2 Contour C in the complex z plane enclosing singularities at $z_1, \ldots, z_N$. The contours labeled γ_i for $i = 1, \ldots, N$ surround their respective singularities.

The residue is just the coefficient c_{-1} in the Laurent expansion

$$f(z) = \sum_{n=-\infty}^{\infty} c_n(z - z_k)^n . \tag{7.15}$$

An alternative expression (but following the same principle) is

$$\text{Res}_{z=z_k} f(z) = \lim_{z \to z_k} \frac{1}{(m-1)!} \frac{d^{m-1}}{dz^{m-1}} [(z - z_k)^m f(z)] , \tag{7.16}$$

where z_k is a pole of order m of $f(z)$.

Case 1. For contour r (shown in Figure 7.1), $G = 0$ when $z^0 < 0$ because $\text{Im}(k_0) > 0$ and C encircles no singularities. When $z^0 > 0$, however,

$$\oint_{C_r} dk_0 \frac{\exp\{-ik_0(x^0 - x'^0)\}}{k_0^2 - \mathbf{k}^2} = -2\pi i \, \text{Res} \left(\frac{\exp\{-ik_0(x^0 - x'^0)\}}{k_0^2 - \mathbf{k}^2} \right) , \tag{7.17}$$

the negative sign coming from the clockwise sense of C when $\text{Im}(k_0) < 0$. Evaluating the residue is simple when we factor the denominator, $k_0^2 - \mathbf{k}^2 =$

$(k_0 + |\mathbf{k}|)(k_0 - |\mathbf{k}|)$, for then

$$\oint_{C_r} dk_0 \frac{\exp\{-ik_0(x^0 - x'^0)\}}{k_0^2 - \mathbf{k}^2} = -2\pi i \left\{ \lim_{k_0 \to |\mathbf{k}|} \frac{\exp\{-ik_0(x^0 - x'^0)\}}{(k_0 + |\mathbf{k}|)} \right.$$

$$\left. + \lim_{k_0 \to -|\mathbf{k}|} \frac{\exp\{-ik_0(x^0 - x'^0)\}}{(k_0 - |\mathbf{k}|)} \right\} , \quad (7.18)$$

or

$$\oint_{C_r} dk_0 \frac{\exp\{-ik_0(x^0 - x'^0)\}}{k_0^2 - \mathbf{k}^2} = -\frac{2\pi}{|\mathbf{k}|} \sin[|\mathbf{k}|(x^0 - x'^0)] . \quad (7.19)$$

Thus,

$$G_r(x - x') = 4\pi \frac{\Theta(z^0)}{(2\pi)^3} \int d^3|\mathbf{k}| \exp\{i\mathbf{k} \cdot (\mathbf{x} - \mathbf{x}')\} \frac{\sin[|\mathbf{k}|(x^0 - x'^0)]}{|\mathbf{k}|} , \quad (7.20)$$

where $\Theta(z^0)$ is the Heaviside function defined to be $+1$ when the argument $z^0 > 0$ and zero otherwise. Let us now choose the axes so that

$$\mathbf{x} - \mathbf{x}' = (0, 0, x^3 - x'^3) , \quad (7.21)$$

which we are certainly free to do because we are integrating over all angles of $\mathbf{k}$. Then, in polar coordinates,

$$G_r(x - x') = 4\pi \frac{\Theta(x^0 - x'^0)}{(2\pi)^3} \int_0^\infty |\mathbf{k}|^2 d|\mathbf{k}| \sin\theta \, d\theta \, d\phi \, \exp\{i|\mathbf{k}||\mathbf{x} - \mathbf{x}'| \cos\theta\}$$

$$\times \frac{\sin[|\mathbf{k}|(x^0 - x'^0)]}{|\mathbf{k}|} . \quad (7.22)$$

The integrals over θ and ϕ are trivial and lead to the result

$$G_r(x - x') = \frac{8\pi \Theta(x^0 - x'^0)}{(2\pi)^2 |\mathbf{x} - \mathbf{x}'|} \int_0^\infty d|\mathbf{k}| \sin(|\mathbf{k}|[x^0 - x'^0]) \sin(|\mathbf{k}||\mathbf{x} - \mathbf{x}'|) . \quad (7.23)$$

In addition, putting $\sin\theta \equiv (\exp[i\theta] - \exp[-i\theta])/2i$ and noting that the integrand is an even function, we get

$$G_r(x - x') = \frac{\Theta(x^0 - x'^0)}{2\pi |\mathbf{x} - \mathbf{x}'|} \int_{-\infty}^\infty d|\mathbf{k}| \left[\exp\{i|\mathbf{k}|(x^0 - x'^0 - |\mathbf{x} - \mathbf{x}'|)\} \right.$$

$$\left. - \exp\{i|\mathbf{k}|(x^0 - x'^0 + |\mathbf{x} - \mathbf{x}'|)\} \right] . \quad (7.24)$$

Since $(x^0 - x'^0) > 0$ and $|\mathbf{x} - \mathbf{x}'|$ is positive definite, the argument of the second exponential is never zero, and since the integral is a delta function, it therefore does not contribute to the Green function. That is,

$$G_r(x - x') = \frac{\Theta(x^0 - x'^0)}{|\mathbf{x} - \mathbf{x}'|} \delta(x^0 - x'^0 - |\mathbf{x} - \mathbf{x}'|) . \tag{7.25}$$

This recovers the *retarded* (or causal) Green function because the source-point time x'^0 is always earlier than the observation-point time x^0 (compare with Equation [3.23]). The sequence followed to produce the potential A_{in}^μ is shown schematically in Figure 7.3.

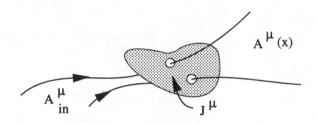

Figure 7.3 An incoming electromagnetic wave A_{in}^μ (known at $x^0 \to -\infty$) interacts with a localized source J^μ to produce the general retarded solution $A^\mu(x)$.

Case 2. A similar calculation for contour a in Figure 7.1 gives $G = 0$ when $z^0 > 0$ because now $\text{Im}(k_0) < 0$ and C encircles no singularities. For $z^0 < 0$, however,

$$G_a(x - x') = \frac{\Theta(-[x^0 - x'^0])}{|\mathbf{x} - \mathbf{x}'|} \delta(x^0 - x'^0 + |\mathbf{x} - \mathbf{x}'|) . \tag{7.26}$$

This is the *advanced* Green function because x'^0 is always later than x^0.

With the retarded and advanced Green functions in hand, we may now complete this section by writing down the solution to the four-dimensional wave equation (7.1). Consider first the case where we have a known incoming electromagnetic wave A_{in}^μ. Based on our previous experience with Equations (3.26) and (3.27), we expect that at any later time x^0, the potential is given by

$$A^\mu(x) = A_{\text{in}}^\mu(x) + \frac{1}{c} \int d^4x' \, G_r(x - x') J^\mu(x') . \tag{7.27}$$

To see this, we note first that $\partial_\alpha \partial^\alpha A_{\text{in}}^\mu(x) = 0$ since A_{in}^μ was produced at some spacetime point $x_{\text{in}} \neq x$. That is, any local J^μ does not contribute to A_{in}^μ. Second, we also know that

$$\partial_\alpha \partial^\alpha \frac{1}{c} \int d^4x'\, G_r(x - x')\, J^\mu(x') = \frac{1}{c} \int d^4x'\, \partial_\alpha \partial^\alpha\, G_r(x - x')\, J^\mu(x')$$

$$= \frac{4\pi}{c} \int d^4x'\, \delta^4(x - x')\, J^\mu(x') = \frac{4\pi}{c}\, J^\mu(x) \, . \tag{7.28}$$

And so, $\partial_\alpha \partial^\alpha A^\mu(x) = 4\pi J^\mu(x)/c$, as required.

If we instead knew the outgoing electromagnetic wave $A_{\text{out}}^\mu(x)$, then we could write

$$A^\mu(x) = A_{\text{out}}^\mu(x) + \frac{1}{c} \int d^4x'\, G_a(x - x')\, J^\mu(x') \, , \tag{7.29}$$

where x'^0 inside the integral must be restricted to values greater than x^0. Clearly, this potential also satisfies the wave equation, since $\partial_\alpha \partial^\alpha A_{\text{out}}^\mu(x) = 0$. As before, writing the potential this way, we are subtracting the contribution to $A_{\text{out}}^\mu(x)$ from sources at $x^0 < x'^0$.

7.2 LIÉNARD-WIECHERT POTENTIALS AND FIELDS FOR A POINT CHARGE

In a given inertial frame K, a point particle's charge and current densities are, respectively,

$$\rho(\mathbf{x}, t) = q\, \delta^3(\mathbf{x} - \mathbf{x}_q[t]) \, , \tag{7.30}$$

$$\mathbf{J}(\mathbf{x}, t) = q\mathbf{v}(t)\, \delta^3(\mathbf{x} - \mathbf{x}_q[t]) \, , \tag{7.31}$$

where $\mathbf{x}_q(t)$ is its trajectory, and by definition, $\mathbf{v}(t) \equiv d\mathbf{x}_q/dt$. In the spirit of our 4D reformulation of radiation theory, we write these in covariant form as follows:

$$J^\mu(x) = qc \int d\tau\, u^\mu(\tau)\, \delta^4(x - x_q[\tau]) \, , \tag{7.32}$$

where the proper time τ parameterizes the trajectory and $u^\mu(\tau) \equiv dx_q^\mu(\tau)/d\tau$ is the particle's 4-velocity defined in Equation (5.125). To confirm that J^μ reduces to the correct frame-dependent densities, we use the property of the Dirac delta function expressed in Equation (4.154). Putting

$$f(\tau) \equiv x^0 - x_q^0 \, , \tag{7.33}$$

with

$$x^0 - x_q^0(\tau_0) = 0 \,, \tag{7.34}$$

we see that

$$\frac{df}{d\tau} = -\frac{dx_q^0}{d\tau} \,. \tag{7.35}$$

Thus, according to Equation (4.154), we should have

$$J^\mu(x) = qc\, u^\mu(\tau_0)\, \delta^3(\mathbf{x} - \mathbf{x}_q[\tau_0]) \left| \frac{dx_q^0}{d\tau} \right|_{\tau_0}^{-1} \,. \tag{7.36}$$

But

$$\left| \frac{dx_q^0}{d\tau} \right|_{\tau_0} = \frac{dx_q^0}{dt} \left| \frac{dt}{d\tau} \right|_{\tau_0} \,, \tag{7.37}$$

and

$$u^\mu(\tau_0) = \left(\frac{dx^\mu}{d\tau} \right)_{\tau_0} = \frac{dx^\mu}{dt} \left| \frac{dt}{d\tau} \right|_{\tau_0} \,. \tag{7.38}$$

Therefore,

$$J^\mu(x) = q\delta^3(\mathbf{x} - \mathbf{x}_q[\tau_0]) \frac{dx^\mu}{dt} \,, \tag{7.39}$$

which is the required form.

Returning now to the general solution for $A^\mu(x)$ in Equation (7.27), and assuming the absence of an incoming wave (i.e., the entire potential is due to the action of the charge), we see that

$$A^\mu(x) = \frac{1}{c} \int d^4x' \, \frac{\Theta(x^0 - x'^0)}{|\mathbf{x} - \mathbf{x}'|} \, \delta(x^0 - x'^0 - |\mathbf{x} - \mathbf{x}'|) \, J^\mu(x') \,. \tag{7.40}$$

One way to proceed from here is to use the identity

$$\begin{aligned}
\delta\left\{(x - x')^2\right\} &= \delta\left\{(x^0 - x'^0)^2 - |\mathbf{x} - \mathbf{x}'|^2\right\} \\
&= \delta\left\{(x^0 - x'^0 - |\mathbf{x} - \mathbf{x}'|)(x^0 - x'^0 + |\mathbf{x} - \mathbf{x}'|)\right\} \\
&= \frac{1}{2|\mathbf{x} - \mathbf{x}'|} \left\{\delta(x^0 - x'^0 - |\mathbf{x} - \mathbf{x}'|) \right. \\
&\qquad \left. + \, \delta(x^0 - x'^0 + |\mathbf{x} - \mathbf{x}'|)\right\} \,.
\end{aligned} \tag{7.41}$$

Since the Θ function selects only one of these two terms by virtue of the fact that both $(x^0 - x'^0)$ and $|\mathbf{x} - \mathbf{x}'|$ are positive for the retarded case, we may thus write

$$G_r(x - x') = 2\Theta(x^0 - x'^0)\, \delta\{(x - x')^2\} \,. \tag{7.42}$$

Clearly, the advanced case is similar, except that x'^0 and x^0 are reversed:

$$G_a(x - x') = 2\Theta(x'^0 - x^0)\,\delta\{(x - x')^2\}\,. \qquad (7.43)$$

And so rearranging the integrand in Equation (7.40), we arrive at

$$A^\mu(x) = 2q \int d^4x'\,\Theta(x^0 - x'^0)\,\delta\{(x - x')^2\} \int d\tau\,u^\mu(\tau)\,\delta^4\{x' - x_q(\tau)\}\,. \quad (7.44)$$

Let us rewrite this in the form

$$A^\mu(x) = 2q \int d\tau \int d^4x'\,\Theta(x^0 - x'^0)\,\delta\{(x - x')^2\}\,u^\mu(\tau)\,\delta^4\{x' - x_q(\tau)\}\,, \quad (7.45)$$

suggesting the order in which we will carry out the integrations, motivated by the fact that the argument of the second delta function is linear in the integration variable x'. Thus,

$$A^\mu(x) = 2q \int d\tau\,\Theta\{x^0 - x_q^0(\tau)\}\,\delta\{[x - x_q(\tau)]^2\}\,u^\mu(\tau)\,. \qquad (7.46)$$

The argument of the remaining delta function is not linear in τ, and we must again use the identity expressed in Equation (4.154), for which we need

$$
\begin{aligned}
f(\tau) &\equiv [x - x_q(\tau)]^2 \\
&= \{x^0 - x_q^0(\tau)\}^2 - \{\mathbf{x} - \mathbf{x}_q(\tau)\}^2 \\
&= \{x^0 - x_q^0(\tau) + |\mathbf{x} - \mathbf{x}_q(\tau)|\} \times \{x^0 - x_q^0(\tau) - |\mathbf{x} - \mathbf{x}_q(\tau)|\}\,. \quad (7.47)
\end{aligned}
$$

The zeros of $f(\tau)$ are given by the *light-cone conditions* (Figure 7.4)

$$x^0 - x_q^0(\tau_0^-) = |\mathbf{x} - \mathbf{x}_q(\tau_0^-)|\,, \qquad (7.48)$$

and

$$x^0 - x_q^0(\tau_0^+) = -|\mathbf{x} - \mathbf{x}_q(\tau_0^+)|\,. \qquad (7.49)$$

Now,

$$\frac{df}{d\tau} = 2[x^\mu - x_q^\mu(\tau)]\,u_\mu(\tau)\,, \qquad (7.50)$$

so that

$$\left.\frac{df}{d\tau}\right|_{\tau_0} = 2[x^\mu - x_q^\mu(\tau_0)]\,u_\mu(\tau_0)\,. \qquad (7.51)$$

Putting all this together gives us the solution for $A^\mu(x)$:

$$A^\mu(x) = \left\{\frac{q u^\mu(\tau)}{[x^\alpha - x_q^\alpha(\tau)]\,u_\alpha(\tau)}\right\}_{\tau = \tau_0^-}\,. \qquad (7.52)$$

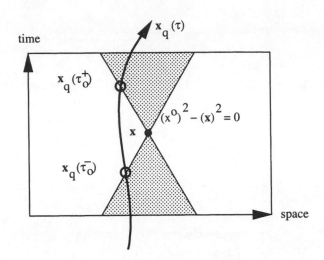

Figure 7.4 The world line $\mathbf{x}_q(\tau)$ of a charge q shown in a reference frame in which an observer is measuring the electromagnetic field at the spacetime point $\mathbf{x}$. Extending the light cones into the future and the past from $\mathbf{x}$ shows that there are two locations where the particle's trajectory intersects the light cones, since the particle's path cannot ever have a gradient shallower than that of light, which would require a velocity greater than c. The two intersection points are the light-cone conditions expressed in Equations (7.48) and (7.49).

Note that although two solutions τ_0^- and τ_0^+ are possible, only one (the retarded one) is allowed by Θ in the Green function. As we would expect based on our earlier experience in Chapter 4, the charge contributes to the potential $A^\mu(x)$ only at the retarded time τ_0^-, defined by the light-cone condition

$$\left\{ x - x_q(\tau_0^-) \right\}^2 = 0 \, . \tag{7.53}$$

From here, it is a straightforward task to reduce $A^\mu(x)$ to its noncovariant form, constituting the Liénard-Wiechert potentials that we derived earlier in § 4.6.1. Let us write

$$
\begin{aligned}
u_\alpha(x^\alpha - x_q^\alpha[\tau_0]) &= u_0 \, (x^0 - x_q^0[\tau_0]) - \mathbf{u} \cdot (\mathbf{x} - \mathbf{x}_q[\tau_0]) \\
&= \gamma c |\mathbf{x} - \mathbf{x}_q(\tau_0)| - \gamma \mathbf{v} \cdot (\mathbf{x} - \mathbf{x}_q[\tau_0]) \, , \tag{7.54}
\end{aligned}
$$

where we have made use of the retarded time condition in Equation (7.53). That is,

$$u_\alpha(x^\alpha - x_q^\alpha[\tau_0]) = \gamma c |\mathbf{x} - \mathbf{x}_q(\tau_0)| (1 - \vec{\beta} \cdot \hat{n}) \, , \tag{7.55}$$

where

$$\hat{n} \equiv \frac{(\mathbf{x} - \mathbf{x}_q[\tau_0])}{|\mathbf{x} - \mathbf{x}_q(\tau_0)|} \, . \tag{7.56}$$

Thus, in a frame for which the potential decomposition is $A^\mu = (\Phi, \mathbf{A})$, (7.52) becomes

$$\Phi(\mathbf{x}, t) = \left[\frac{q}{(1 - \vec{\beta} \cdot \hat{n})|\mathbf{x} - \mathbf{x}_q|} \right]_{\text{ret}} , \qquad (7.57)$$

and

$$\mathbf{A}(\mathbf{x}, t) = \left[\frac{q\vec{\beta}}{(1 - \vec{\beta} \cdot \hat{n})|\mathbf{x} - \mathbf{x}_q|} \right]_{\text{ret}} , \qquad (7.58)$$

where "ret" means that the quantities inside the bracket are to be evaluated at the retarded time τ_0^- as given in Equation (7.48). In this frame, the fields are correctly given by (1.35) and (1.37), and so we recover the solutions (4.166) and (4.167), except that we here write $\mathbf{E}$ in a slightly different form using the Lorentz factor γ:

$$\mathbf{E}(\mathbf{x}, t) = q \left[\frac{\hat{n} - \vec{\beta}}{\gamma^2 (1 - \vec{\beta} \cdot \hat{n})^3 R^2} \right]_{\text{ret}} + \frac{q}{c} \left[\frac{\hat{n} \times \{(\hat{n} - \vec{\beta}) \times \dot{\vec{\beta}}\}}{(1 - \vec{\beta} \cdot \hat{n})^3 R} \right]_{\text{ret}} , \qquad (7.59)$$

where $R \equiv |\mathbf{x} - \mathbf{x}(\tau_0)|$ and $\hat{n} \equiv \mathbf{R}/R$. We remind ourselves that the first term falls off as R^{-2} and represents the near (or "velocity") component of $\mathbf{E}$, whereas the second term is the far (or "acceleration") field that falls off as R^{-1}. The acceleration fields are transverse to the radius vector.

7.3 ANGULAR DISTRIBUTION OF THE EMITTED RADIATION

When the particle motion is nonrelativistic, the angular distribution of its emitted field is simply the familiar $\sin^2 \theta$ behavior of dipole radiation, where θ is measured relative to the direction of acceleration. However, when $\beta \to 1$, the acceleration field depends on $\mathbf{v}$ as well as $\mathbf{a}$, and so the distribution is not so simple. We can learn more about the radiation pattern in general by looking at the Poynting flux, which must here be written in terms of the retarded fields:

$$\mathbf{S}_{\text{ret}} = \frac{c}{4\pi} \left[\mathbf{E}_{\text{rad}} \times \mathbf{B}_{\text{rad}} \right]_{\text{ret}} , \qquad (7.60)$$

where

$$\mathbf{E}_{\text{rad}} = \frac{q}{c} \left[\frac{\hat{n} \times \{(\hat{n} - \vec{\beta}) \times \dot{\vec{\beta}}\}}{(1 - \vec{\beta} \cdot \hat{n})^3 R} \right]_{\text{ret}} , \qquad (7.61)$$

and
$$\mathbf{B}_{\text{rad}} = \hat{n} \times \mathbf{E}_{\text{rad}} \ . \tag{7.62}$$

In terms of the particle's dynamical quantities (Figure 7.5),

$$|\mathbf{S}|_{\text{ret}} = \frac{q^2}{4\pi c} \left\{ \frac{1}{R^2} \left| \frac{\hat{n} \times [(\hat{n} - \vec{\beta}) \times \dot{\vec{\beta}}]}{(1 - \vec{\beta} \cdot \hat{n})^3} \right|^2 \right\}_{\text{ret}} \ . \tag{7.63}$$

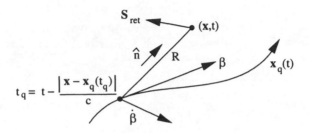

Figure 7.5 The vectors used to calculate the Poynting flux **S** for the radiation field of a particle moving along its trajectory $\mathbf{x}_q(t)$.

But before we start using **S** to analyze the angular distribution of the *emitted* radiation, let's make sure we understand exactly what this quantity represents. Clearly, it must be the energy per unit area per unit time passing by the observation point. However, an interval of time Δt for the observer is not the same as an interval $\Delta t'$ for the emitter. Noting that the solid angle subtended by an element of area $\hat{n}\Delta A$ a distance R from the source is $\Delta\Omega = \Delta A/R^2$, we infer that the power *radiated* into $\Delta\Omega$ is

$$\Delta P(t') = \Delta\Omega \, R^2 |\mathbf{S}_{\text{ret}} \cdot \hat{n}| \frac{dt}{dt'} \ , \tag{7.64}$$

implying a radiated power per unit solid angle

$$\frac{dP(t')}{d\Omega} = R^2 |\mathbf{S}_{\text{ret}} \cdot \hat{n}| \frac{dt}{dt'} \ . \tag{7.65}$$

We found that the retarded time can be expressed in terms of t and the distance to the source, i.e.,

$$t = t' + \frac{|\mathbf{x} - \mathbf{x}(t')|}{c} \ , \tag{7.66}$$

whereby

$$\frac{dt}{dt'} = 1 + \frac{\{-\dot{x}(t')[x - x(t')] - \dot{y}(t')[y - y(t')] - \dot{z}(t')[z - z(t')]\}}{c|\mathbf{x} - \mathbf{x}(t')|}$$

$$= (1 - \vec{\beta} \cdot \hat{n})|_{\text{ret}} .$$ (7.67)

Thus, we arrive at the very important result

$$\boxed{\frac{dP(t')}{d\Omega} = \frac{q^2}{4\pi c} \frac{[\hat{n} \times \{(\hat{n} - \vec{\beta}) \times \dot{\vec{\beta}}\}]^2}{(1 - \hat{n} \cdot \vec{\beta})^5}\Bigg|_{\text{ret}}} .$$ (7.68)

Example 7.1. Suppose the particle is being accelerated in a direction parallel to its velocity, i.e., $\dot{\vec{\beta}} \propto \vec{\beta}$ (Figure 7.6).

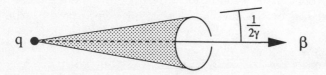

Figure 7.6 The radiation beaming cone for a particle being accelerated in the direction of its motion.

Then,

$$\frac{dP(t')}{d\Omega} = \frac{q^2 \dot{v}^2}{4\pi c^3} \frac{\sin^2 \theta}{(1 - \beta \cos \theta)^5} ,$$ (7.69)

where $\beta \equiv v/c$ and $\sin \theta = \hat{n} \times \dot{\vec{\beta}}/|\dot{\vec{\beta}}|$. The intensity reaches a maximum at some angle θ_m, where

$$\frac{d}{d(\cos \theta)} \left(\frac{dP(t')}{d\Omega} \right) = 0 ,$$ (7.70)

yielding the condition

$$\cos \theta_m = \frac{\sqrt{1 + 15\beta^2} - 1}{3\beta} .$$ (7.71)

When $\beta \to 1$, we find that

$$\theta_m(\beta \to 1) \approx \frac{1}{2\gamma} \ll 1 , \tag{7.72}$$

and we conclude that most of the power is therefore emitted (perhaps beamed is a better word) within a relatively narrow cone with opening angle $\sim 2\theta_m = 1/\gamma$ about $\dot{\vec{\beta}} \propto \vec{\beta}$.

7.4   BREMSSTRAHLUNG RADIATION

As a straightforward application of some of the ideas we have been developing in Chapters 5 and 7, we will now consider the free-free (or bremsstrahlung) radiation due to the acceleration of a charge in the Coulomb field of another charge. A full treatment of this process requires quantum mechanics, since photons of energy comparable to that of the emitting particle can be produced (Heitler 1954). Our discussion here is valid only for a frequency $\omega \ll \gamma mc^2/\hbar$ of the emitted radiation. In addition, quantum mechanical effects manifest themselves when the collision distance shrinks to a value smaller than that of a region specified by the uncertainty principle. For a given incoming particle momentum $p = \gamma mv$, its path may be defined only to within an uncertainty $\Delta b \gtrsim \hbar/p$. Classically, however, a particle with charge e and velocity v can approach another particle with charge Ze to within a distance $b_{\min} \approx Ze^2/(\gamma - 1)mc^2$. Thus, a classical treatment is valid only as long $b_{\min}/\Delta b \gg 1$, which occurs when $2Ze^2/\hbar v \gg 1$, or $v/c \ll (Z/95)$. The classical treatment we consider here is accurate only when these conditions are satisfied.

We proceed by using a technique known as the method of virtual quanta (Williams 1935). Let us consider the collision between a relativistic electron and a heavy ion of charge Ze, which is moving much more slowly as seen in the laboratory frame (Figure 7.7). We shall begin by first moving to a reference frame (the "primed" frame) in which the electron is initially at rest. For simplicity (and without any loss of generality), assume that the ion moves along the x'-axis. Based on our analysis in §§ 4.6.2 and 7.2, we expect that when $|\mathbf{v}'| \to c$, the electrostatic field of the ion is transformed into an essentially transverse wave with $|\mathbf{E}'| \sim |\mathbf{B}'|$, which appears to the electron to be a pulse of electromagnetic radiation. It is this "bunched" field that Compton scatters off the electron to produce the emitted radiation. In effect, relativistic bremsstrahlung can be regarded as the Compton scattering of the *virtual quanta* of the ion's electrostatic field as seen in the electron's frame. We shall see how this works in practice below.

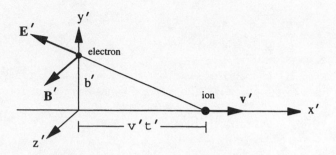

Figure 7.7 The collision between an electron (initially at rest in the primed frame) and an ion moving along the x'-axis with velocity $\mathbf{v}'$. $\mathbf{E}'$ and $\mathbf{B}'$ are the electric and magnetic fields due to the ion at the position of the electron.

In the ion's rest frame, which we shall refer to as the double-primed frame, the electric and magnetic fields are simply given as

$$E''_x = Ze\,x''/r''^3 \qquad\qquad B''_x = 0\,,$$

$$E''_y = Ze\,y''/r''^3 \qquad\qquad B''_y = 0\,,$$

$$E''_z = Ze\,z''/r''^3 \qquad\qquad B''_z = 0\,, \qquad (7.73)$$

where $r'' \equiv \sqrt{x''^2 + y''^2 + z''^2}$. Thus, according to the field transformation Equations (5.131) and (5.132) with the appropriate inversions and change of boosting direction, the electron "sees" the field components

$$E'_x = E''_x = Ze\,x''/r''^3\,,$$

$$E'_y = \gamma(v')\left(E''_y + \frac{v'}{c}\cdot 0\right) = \gamma(v')\,Ze\,y''/r''^3\,,$$

$$E'_z = \gamma(v')\left(E''_z - \frac{v'}{c}\cdot 0\right) = \gamma(v')\,Ze\,z''/r''^3\,, \qquad (7.74)$$

and similarly for $\mathbf{B}'$, where $\gamma(v')$ is the Lorentz factor defined in Equation (5.14), but here in terms of the ion's velocity v' as seen by the electron. However, we must also remember to transform the coordinates appearing in Equation (7.74):

$$x'' = \gamma(v')\,(x' - v't')\,,$$

$$y'' = y'\,,$$

$$z'' = z' \, ,$$

$$r'' = \left\{ \gamma^2(v')(x' - v't')^2 + y'^2 + z'^2 \right\}^{1/2} \, . \tag{7.75}$$

If we now choose the coordinates such that $t' = 0$ when the charges have their closest approach (see Figure 7.7), these transformations reduce to

$$y'' \;\; = \;\; y' = b' \, ,$$

$$z'' \;\; = \;\; z' = 0 \, ,$$

$$x'' \;\; = \;\; -\gamma(v') \, v't' \, , \tag{7.76}$$

and so the electric and magnetic field components acting on the electron in its own rest frame are given by the expressions

$$E'_x = \frac{-Ze \, \gamma(v') \, v't'}{[\gamma^2(v')v'^2 t'^2 + b'^2]^{3/2}} \qquad B'_x = 0 \, ,$$

$$E'_y = \frac{Ze \, \gamma(v') \, b'}{[\gamma^2(v')v'^2 t'^2 + b'^2]^{3/2}} \qquad B'_y = 0 \, ,$$

$$E'_z = 0 \qquad\qquad\qquad B'_z = \frac{v'}{c} \, E'_y \, . \tag{7.77}$$

Clearly, the fields seen by the electron are strongest when $\gamma(v') \, v't' \lesssim b'$, that is, for times

$$t' \lesssim \frac{b'}{\gamma(v') \, v'} \, . \tag{7.78}$$

This means that the fields are concentrated in a plane transverse to the ion's velocity, within an angle θ', where

$$\tan \theta' \approx \frac{v't'}{b'} \approx \frac{1}{\gamma(v')} \, . \tag{7.79}$$

Since $\gamma(v') \gg 1$, this reduces to the condition $\theta' \approx 1/\gamma(v')$, and we see why the field of a highly relativistic charge appears to be a pulse of radiation traveling in the same direction as the particle, confined to a narrow transverse region, as shown schematically in Figure 7.8.

To find the spectrum of this radiation pulse, we use the magnitude of the Poynting vector

$$|\mathbf{S}'| = \frac{dW'}{dt' \, dA'} = \frac{c}{4\pi} \, \mathbf{E}'^2(t') \, , \tag{7.80}$$

representing the energy flux per unit time per unit area. The total energy per unit area in the pulse is

$$\frac{dW'}{dA'} = \frac{c}{4\pi} \int_{-\infty}^{\infty} \mathbf{E}'^2(t') \, dt' \, . \tag{7.81}$$

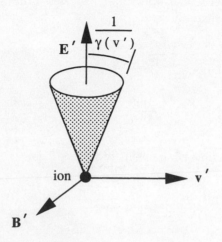

Figure 7.8 The field of a highly relativistic ion as seen in the electron's rest (i.e., primed) frame. Using the coordinate system defined in Figure 7.7, we see that $\mathbf{E}'$ is concentrated in the y'-direction whereas $\mathbf{B}'$ points along the z'-axis.

But now writing the Fourier transform of $E'(t')$ as

$$\hat{E}'(\omega') = \frac{1}{2\pi} \int_{-\infty}^{\infty} E'(t') \, \exp(i\omega' t') \, dt' \, , \tag{7.82}$$

we know from Parseval's theorem for Fourier transforms that

$$\int_{-\infty}^{\infty} E'^2(t') \, dt' = 2\pi \int_{-\infty}^{\infty} |\hat{E}'(\omega')|^2 \, d\omega' \, . \tag{7.83}$$

Since $E'(t')$ is real,

$$\hat{E}'(-\omega') = \frac{1}{2\pi} \int_{-\infty}^{\infty} E'(t') \, \exp(-i\omega' t') \, dt' = \hat{E}'^*(\omega') \, , \tag{7.84}$$

so that

$$|\hat{E}'(\omega')|^2 = |\hat{E}'(-\omega')|^2 \ , \tag{7.85}$$

and therefore,

$$\int_{-\infty}^{\infty} E'^2(t') \ dt' = 4\pi \int_0^{\infty} |\hat{E}'(\omega')|^2 \ d\omega' \ . \tag{7.86}$$

We do this in order to end up with an expression that involves only physically meaningful frequencies. Thus,

$$\frac{dW'}{dA'} = c \int_0^{\infty} |\hat{E}'(\omega')|^2 \ d\omega' \ , \tag{7.87}$$

and so

$$\frac{dW'}{dA' \ d\omega'} = c|\hat{E}'(\omega')|^2 \ . \tag{7.88}$$

Ignoring the negligible components of $\mathbf{E}'$ (see Equation [7.77]), we have in this application

$$\hat{E}'(\omega') = \frac{1}{2\pi} \int_{-\infty}^{\infty} E_y'(t') \ \exp(i\omega't') \ dt' \ , \tag{7.89}$$

or more explicitly,

$$\hat{E}'(\omega') = \frac{Ze \, \gamma(v') \, b'}{2\pi} \int_{-\infty}^{\infty} \frac{\exp(i\omega't') \ dt'}{[\gamma^2(v') \, v'^2 t'^2 + b'^2]^{3/2}} \ , \tag{7.90}$$

which can be evaluated in terms of the modified Bessel function K_1 of order one:

$$\hat{E}'(\omega') = \frac{Ze}{\pi b'v'} \left(\frac{b'\omega'}{\gamma(v')v'} \right) K_1 \left(\frac{b'\omega'}{\gamma(v')v'} \right) \ . \tag{7.91}$$

Thus, the spectrum is

$$\frac{dW'}{dA' \ d\omega'} = c|\hat{E}'(\omega')|^2 = \frac{(Ze)^2 c}{\pi^2 b'^2 v'^2} \left(\frac{b'\omega'}{\gamma(v')v'} \right)^2 K_1^2 \left(\frac{b'\omega'}{\gamma(v')v'} \right) \ . \tag{7.92}$$

Note that this frequency distribution starts to cut off when

$$\omega' > \gamma(v')v'/b' \ , \tag{7.93}$$

which is consistent with the fact that the pulse is confined roughly to a time interval of order $t' \sim b'/\gamma(v')v'$. Thus, the characteristic cutoff frequency should be

$$\omega' \sim \frac{1}{t'} \sim \frac{\gamma(v')v'}{b'} \ , \tag{7.94}$$

which is the value we infer from the complete expression (7.92). Of particular interest is the situation where $v' \to c$. According to our discussion at the beginning of this subsection, this can occur for sufficiently large values of Z. In that case,

$$\frac{dW'}{dA'\,d\omega'} = \frac{(Ze)^2}{\pi^2 b'^2 c}\left(\frac{b'\omega'}{\gamma(v')c}\right)^2 K_1^2\left(\frac{b'\omega'}{\gamma(v')c}\right). \tag{7.95}$$

This is the photon spectrum, in the form of a pulse, that scatters off the electron to produce the bremsstrahlung radiation we see in the laboratory frame. To simplify matters, let us consider the low-frequency domain (specifically, the region $\hbar\omega'/2\pi \lesssim m_e c^2$), in which the scattering of the virtual quanta by the electron occurs in the Thomson limit with a constant cross section σ_T. The scattered radiation is then

$$\left.\frac{dW'}{d\omega'}\right|_{\text{scatt}} = \sigma_T \frac{dW'}{dA'\,d\omega'}. \tag{7.96}$$

And in the final step, when we wish to transform this to the laboratory frame where the radiation is actually measured, we note that both the energy W and the frequency ω transform as the 0-components of 4-vectors (i.e., they transform identically), so that $dW/d\omega$ is invariant. The scattered virtual radiation, or what we would otherwise call the relativistic bremsstrahlung emission, therefore has the following distribution in frequency as seen by an observer in the laboratory:

$$\left.\frac{dW}{d\omega}\right|_{\text{lab}} = \sigma_T \frac{(Ze)^2}{\pi^2 b'^2 c}\left(\frac{b'\omega'}{\gamma(v')c}\right)^2 K_1^2\left(\frac{b'\omega'}{\gamma(v')c}\right). \tag{7.97}$$

Earlier, we mentioned that the ion is moving slowly in our frame. We haven't actually used this yet, but we can now see why this restriction simplifies matters somewhat, for v' is then almost equal to the electron's velocity v in the laboratory. In other words, the electron's velocity is by far the dominant contributor to the relativistic effects during the transformation, and so we can put $\gamma(v') \approx \gamma(v)$. In addition, we can put $b = b'$ since the impact parameter is measured perpendicular to the relative direction of motion between the electron and the proton. The only remaining frame-dependent quantity appearing in Equation (7.97) is the frequency, which we may transform to its laboratory frame counterpart using Equation (5.31), though actually in this case it's easier to use the reverse transformation

$$\omega = \gamma(v)\omega'(1 + \beta\cos\theta'), \tag{7.98}$$

where θ' is the scattering angle in the electron's rest frame. In the Thomson limit, an observer in this frame sees scattered radiation with front-back symmetry, meaning that when we average out the angle dependence in Equation (7.98), $\omega \sim \gamma(v)\omega'$. Thus, with the final substitution

$$\sigma_\text{T} \equiv \frac{8\pi}{3} \frac{e^4}{m_e^2 c^4} \,, \tag{7.99}$$

we get the laboratory frame bremsstrahlung emissivity (Figure 7.9)

$$\frac{dW}{d\omega} = \frac{8Z^2 e^6}{3\pi b^2 c^5 m_e^2} \left(\frac{b\omega}{\gamma^2 c}\right)^2 K_1^2 \left(\frac{b\omega}{\gamma^2 c}\right) \,. \tag{7.100}$$

Asymptotically, we know that

$$K_1(x) \sim \frac{1}{x} \,, \qquad (x \ll 1) \,, \tag{7.101}$$

and

$$K_1(x) \sim \left(\frac{\pi}{2x}\right)^{1/2} \exp(-x) \,, \qquad (x \gg 1) \,. \tag{7.102}$$

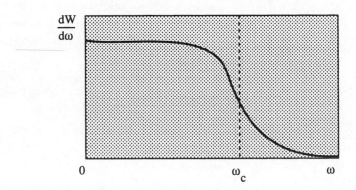

Figure 7.9 Sketch of the bremsstrahlung spectrum, which is independent of frequency for low values of ω, and falls off exponentially for $\omega > \omega_c$, where $\omega_c \equiv \gamma^2 c/b$.

Thus, at low frequencies (i.e., $\omega \ll \omega_c \equiv \gamma^2 c/b$), $dW/d\omega$ is independent of ω. However, at large frequencies (i.e., $\omega \gg \omega_c$),

$$\frac{dW}{d\omega} \sim \frac{\omega}{b} \exp\left(\frac{-2b\omega}{\gamma^2 c}\right) \,, \tag{7.103}$$

and so the spectrum falls off exponentially. The overall distribution is shown schematically in Figure 7.9.

This type of particle emissivity occurs quite frequently in nature and is extremely useful in providing us with a means to infer the underlying physical conditions at the source. High-temperature plasmas, such as those encountered in confined fusion systems, or near the surface of very compact astrophysical objects, are particularly good examples of this. In a hot gaseous environment, the average particle energy is given in terms of the plasma temperature T according to $\langle E \rangle \approx 3kT/2$. Thus, when a particle with this energy is incident on a scattering center, its distance of closest approach b corresponds to the situation where all of this kinetic energy is converted into Coulomb potential energy (Ze^2/b), so that in general, $b \sim T^{-1}$. Since in addition $\omega_c = \omega_c(b, T)$, it is clear that ω_c therefore effectively depends only on the temperature, and the location of a "knee" in the observed spectrum immediately gives us a measure of the particle energy through the inferred value of T.

7.5   RADIATIVE MOTIONS OF A POINT CHARGE

What happens to a charge while it is radiating? Although this may sound like a philosophical inquiry, it is in fact one of the most intriguing questions we can ask in electrodynamics. We have avoided thinking about this too hard all along with the justification that if we stay in the appropriate particle energy regimes and consider only sufficiently low-frequency fields, then the particle motion can be described as if the process of radiative emission has no effect on the sources themselves. In other words, we have assumed that while they are radiating, the particles undergo only minimal changes to their energy, momentum, and angular momentum as a result of this process. But surely something must be happening to them, since the fields do in fact carry dynamical luggage away from the charges, as we showed in § 3.3. And yet when we measure the mass of the electron before and after it has radiated, it always has the same value. With our development of a relativistically correct theory, we are now in a position to stray away from our previously self-imposed restrictive energy ranges and begin to explore what our presumably correct field and dynamics equations tell us about the process of radiation in more extreme situations. Our hope is that by stretching our analysis in this way, we will start to see how the charges react to the creation of a field and thereby to understand the process of emission itself.

As we saw in § 4.6.2, the Poynting vector representing the radiative flux emitted by an accelerated charge has a magnitude given by Equation (4.176). In terms of the elemental solid angle $d\Omega \equiv dA/r^2$, where dA is an increment of area, the power radiated by the charge per unit solid angle is $dP/d\Omega = q^2 a^2 \sin^2\theta/4\pi c^3$. Thus, in classical electrodynamics, an accelerated charge q emits radiation at the Larmor rate

$$P = \frac{2q^2}{3c^3}(\dot{\mathbf{v}})^2 \,, \tag{7.104}$$

where $a \equiv |\dot{\mathbf{v}}|$, which is the integral of $dP/d\Omega$ over all solid angles. One often assumes that the particle's trajectory (and hence $\dot{\mathbf{v}}$) are known a priori and so calculating the rate of energy loss P is straightforward, if not completely accurate. A correct treatment of this problem would require the equations describing the accelerated motions of point charges to be modified if the reactions on the motions from their radiations are to be taken into account. Let us add a *radiative reaction force* $\mathbf{F}_{\text{rad}}$ to the Newtonian equation of motion, so that

$$m\dot{\mathbf{v}} = \mathbf{F}_{\text{ext}} + \mathbf{F}_{\text{rad}} \,, \tag{7.105}$$

where $\mathbf{F}_{\text{rad}}$ must be chosen to account for the energy loss given by Equation (7.104). But how do we determine what $\mathbf{F}_{\text{rad}}$ looks like? One of the physical arguments we can make is that energy should be strictly conserved during the interaction, and so over the course of the motion (from time t_1 to t_2) we should have

$$\int_{t_1}^{t_2} \mathbf{F}_{\text{rad}} \cdot \mathbf{v}\, dt = -\frac{2q^2}{3c^3} \int_{t_1}^{t_2} \dot{\mathbf{v}} \cdot \dot{\mathbf{v}}\, dt \,. \tag{7.106}$$

In a sense, this is a way of calculating the average radiative reaction force acting over the time interval $t_2 - t_1$. Now, integrating by parts, we have

$$-\frac{2q^2}{3c^3} \int_{t_1}^{t_2} \dot{\mathbf{v}} \cdot \dot{\mathbf{v}}\, dt = -\frac{2q^2}{3c^3}(\dot{\mathbf{v}} \cdot \mathbf{v})|_{t_1}^{t_2} + \frac{2q^2}{3c^3} \int_{t_1}^{t_2} \ddot{\mathbf{v}} \cdot \mathbf{v}\, dt \,. \tag{7.107}$$

Generally speaking, the acceleration acts over a finite time so that by choosing t_1 and t_2 appropriately, we can impose the condition that $\dot{\mathbf{v}} \to 0$ at the end points of the time interval. When this is the case,

$$\int_{t_1}^{t_2} dt \left\{ \mathbf{F}_{\text{rad}} - \frac{2q^2}{3c^3}\ddot{\mathbf{v}} \right\} \cdot \mathbf{v} = 0 \,, \tag{7.108}$$

and since this should be true regardless of what the particle's velocity profile is, we may identify the radiative reaction force as (Iwanenko and Sokolow 1953)

$$\mathbf{F}_{\text{rad}} = \frac{2q^2}{3c^3}\ddot{\mathbf{v}} \,. \tag{7.109}$$

It follows that the dynamics equation for the radiating particle should then be modified to what is known as the Abraham-Lorentz equation of motion (Lorentz 1915)

$$\boxed{m(\dot{\mathbf{v}} - \tau_c\ddot{\mathbf{v}}) = \mathbf{F}_{\text{ext}}} \; , \tag{7.110}$$

where

$$\tau_c \equiv \frac{2q^2}{3mc^3} \approx 6.3 \times 10^{-24} \left(\frac{m_e}{m}\right)\left(\frac{q}{e}\right)^2 \quad \text{s} \tag{7.111}$$

is roughly the time it takes light to cross the classical electron radius $r_e \equiv e^2/m_e c^2$ when $m = m_e$.

It is clear therefore, and this confirms what we have been assuming all along, that the radiative reaction $m\tau_c\ddot{\mathbf{v}}$ is almost always negligible during accelerations by a force $\mathbf{F}_{\text{ext}}$ that falls within the confines of classical analysis. A force producing acceleration variations violent enough to make $\tau_c\ddot{\mathbf{v}} \sim \dot{\mathbf{v}}$ would have to contain Fourier components with frequencies of the order of $1/\tau_c$. In this regime, the field quanta are γ-rays with energy $h/2\pi\tau_c \approx 100$ MeV. However, electron-positron pair creation sets in at $h\nu \gtrsim 2m_e c^2 \sim 1$ MeV. At these energies, the classical analysis is no longer valid since quantum effects cannot be ignored.

Example 7.2. Although in the classical domain the radiative reaction is small, it is nonetheless nonzero. Let us consider the case of a nonrelativistic charged harmonic oscillator to see what observable effect, if any, we can expect. The nonrelativistic equation of motion we should use is Equation (7.110), with a harmonic external force

$$\mathbf{F}_{\text{ext}} = -k\mathbf{x} \; , \tag{7.112}$$

corresponding to a natural frequency $\omega_0 = \sqrt{k/m}$. The equation of motion for a one-dimensional problem is thus

$$\ddot{x} + \omega_0^2 x = \tau_c \frac{d\ddot{x}}{dt} \; , \tag{7.113}$$

and this is what we must now solve for x. Putting $x \sim e^{i\omega t}$ in this equation yields the dispersion relation $\omega^2 - \omega_0{}^2 - i\omega^3\tau_c = 0$. Fortunately, we know that in most circumstances $\tau_c \ll 1/\omega$ and $\omega \approx \omega_0$, so that our solution should not differ greatly from that in the absence of damping. That is, we expect that

$$x \sim \exp(i\omega_0 t) \; , \tag{7.114}$$

for which

$$\dot{x} \sim i\omega_0 x \,,$$

$$\ddot{x} \sim -\omega_0^2 x \,,$$

and

$$\frac{d\ddot{x}}{dt} \sim -i\omega_0^3 x = -\omega_0^2 \dot{x} \,. \tag{7.115}$$

We can thus rewrite Equation (7.113) as

$$\ddot{x} + \omega_0^2 x \approx -\tau_c \omega_0^2 \dot{x} \,, \tag{7.116}$$

or

$$\dddot{x} + \Gamma \dot{x} + \omega_0^2 x \approx 0 \,, \tag{7.117}$$

where $\Gamma \equiv \tau_c \omega_0^2$ is the "damping constant." Since τ_c is very small, most physical systems will have $\Gamma \ll 1$, and so to lowest order in Γ, the radiatively damped solution is

$$x \approx A \exp(-i\omega_0 t) \exp(-\Gamma t/2) \,, \tag{7.118}$$

as may be verified by direct substitution into Equation (7.117). A is the initial amplitude of the motion. We see that the potential energy $U \propto x^2$ of the oscillator thus decays as $\exp(-\Gamma t)$. So yes, the charge does radiate because it has an accelerated motion, and over time the cost is borne by its potential energy which eventually goes to zero.

Because of radiation damping, the fields emitted by the oscillator cannot be monochromatic, since the motion is changing over time. We can estimate the emitted line width by considering a Fourier expansion of the field, which will contain all the contributing frequency components. Let us put

$$E(\mathbf{x}, t) = E_0(\mathbf{x}) \exp(-i\omega t) \exp(-\Gamma t/2) = \int_{-\infty}^{\infty} \hat{E}_\omega(\mathbf{x}) \exp(-i\omega t) \, d\omega \,, \tag{7.119}$$

where the time dependence of E is suggested by the solution for x. Then, the Fourier component with frequency ω is

$$\hat{E}_\omega(\mathbf{x}) = \frac{E_0(\mathbf{x})}{2\pi} \int_0^\infty \exp(-i\omega_0 t) \exp(-\Gamma t/2) \exp(i\omega t) \, dt$$

$$= \frac{E_0(\mathbf{x})}{2\pi} \frac{1}{i(\omega - \omega_0) - \Gamma/2} \,. \tag{7.120}$$

The radiation intensity I_ω is proportional to $\hat{E}_\omega^2$, so that

$$I_\omega = \frac{I_0 \Gamma}{2\pi} \frac{1}{(\omega - \omega_0)^2 + \Gamma^2/4} \,, \tag{7.121}$$

normalized in such a way that

$$\int_{-\infty}^{\infty} I_\omega \, d\omega = I_0 \qquad (7.122)$$

(see Figure 7.10).

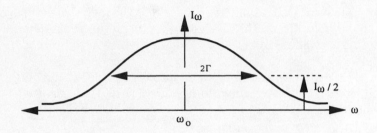

Figure 7.10 Broadening of the emission line from a charged harmonic oscillator, due to the damping effect of the radiative reaction force on the particle. Γ is the half-width of the line at half maximum intensity, and ω_0 is the natural frequency of the oscillator.

The frequency half-width at half-intensity is therefore[20]

$$\Delta\omega_{1/2} \approx \Gamma = \tau_c \, \omega_0^2 \, . \qquad (7.123)$$

By way of example, a charged oscillator with the same parameters as a typical electron in an atom would produce a line with a wavelength half-width $\Delta\lambda_{1/2}/\lambda \sim 10^{-8}$.

---

[20]In addition to a broadening of the line, the radiative reaction also induces a line shift due to the dissipative effect of the emission. However, the classical prediction, which is that the shift is much smaller than $\Delta\omega_{1/2}$ is quite different from the correct quantum mechanical value. The reason for this is that even in the absence of "real" photons, the quantized radiation field is subject to vacuum fluctuations involving the spontaneous creation and annihilation of electron-positron pairs, which act on the charged particle to cause a shift in its energy. This radiative shift, known as the Lamb shift, was first observed in 1947 (Lamb and Retherford 1947).

7.6   RADIATION DAMPING AND THE RELATIVISTIC LORENTZ-DIRAC EQUATION

Let us now break away from the overly restrictive low-energy domain we have been considering and instead use a relativistically correct framework to study the behavior of accelerated charges when $v \to c$. Because the radiation rates are greatly enhanced in this limit, we expect that the radiative damping effects will be appreciable. To do this, we need to generalize the Abraham-Lorentz equation into a covariant form.

Consider the electromagnetic force on a charged particle arising only from an externally applied field. According to Equation (5.129), the covariant electromagnetic force on a charge q is

$$f^\alpha \equiv \frac{dp^\alpha}{d\tau} = \frac{q}{c} F^\alpha{}_\gamma u^\gamma \, , \tag{7.124}$$

and we remind ourselves that the 4-momentum $p^\alpha = mu^\alpha$ is given in terms of the particle's *rest* mass m. Let us be very careful and not make any unnecessary assumptions here. This force equation must preserve the correct balance between the effects on the particle's energy and its momentum from frame to frame, since this is a physical law that all observers must agree on. How do we see that? We know that the individual components of f^α are not conserved, but a scalar formed from the contraction of f^α and another 4-vector is a constant. We have several choices to pick from, but let's take the simplest one available to us, which is the 4-velocity. In effect, what we are doing is to take the "projection" of f^α in the direction of the particle's motion (i.e., u_α), and since this projection is a 4-scalar, it must be invariant from frame to frame. Physically, this amounts to determining the relationship between the power exerted on the charge and the rate of change of its momentum in one frame and then using the known properties of a Lorentz transformation to see what restrictions may be imposed on the particle's mass and velocity. We thus make the contraction (i.e., take the projection)

$$u_\alpha f^\alpha = \frac{q}{c} F^\alpha{}_\gamma u^\gamma u_\alpha = 0 \, , \tag{7.125}$$

where the last equality follows from the antisymmetry of $F^\alpha{}_\gamma$. Though seemingly trivial, this is a very fundamental result, for it says that

$$u_\alpha \frac{dp^\alpha}{d\tau} = \frac{m}{2} \frac{d}{d\tau}(u^\alpha u_\alpha) - (u^\alpha u_\alpha)\frac{dm}{d\tau} = 0 \, , \tag{7.126}$$

or since $u^\alpha u_\alpha = c^2$,

$$\boxed{dm/d\tau = 0} \quad .$$ (7.127)

So the external electromagnetic field induces an interaction that does not change the particle's rest mass, only its momentum $\mathbf{p} = \gamma m \mathbf{v}$. This is important in view of the fact that the particle's own field E_{own} represents an energy distribution and therefore a mass density $E_{\text{own}}^2/4\pi c^2$.

This presents us with somewhat of a dilemma because while f^α may impart kinetic energy K and momentum $\mathbf{p}$ in the right proportions to leave the particle's rest mass unchanged,

$$m^2 c^2 = (K + mc^2)^2/c^2 - \mathbf{p}^2 \, ,$$ (7.128)

the depletions by radiation alone do not! In other words, if we try to accelerate a charge with an electromagnetic field $F^\alpha{}_\gamma$, according to our current set of equations, we give it the right combination of K and $\mathbf{p}$ to keep m constant, but as a result of the induced acceleration, it must then radiate away fractions of these that will not preserve the value of m. This is a consequence of the fact that a charged particle can only give up energy and momentum in proportions consistent with $p^\alpha p_\alpha = m^2 c^2 \neq 0$, whereas the radiation field takes up proportions consistent with a zero photon mass (as we discussed earlier in § 3.3.2). But we know that charges emerge from radiative processes with their total rest mass intact, so there must be some additional reaction—a temporary distortion of the attached self-field—that cancels out the rest-mass changing effects!

Understanding what these distortions are is not trivial, but we can begin by analogy with the nonrelativistic case and argue that no matter what, the radiation process must induce a back reaction on the charge, which we shall call $(dp^\alpha/d\tau)_{\text{rad}}$. Now, in the particle rest (i.e., primed) frame, the radiation is emitted with front-back symmetry, implying that incrementally $d\mathbf{p}' = 0$, so that

$$\left(\frac{d\mathbf{p}'}{d\tau}\right)_{\text{rad}} = 0 \, .$$ (7.129)

However, the charge is losing energy at a rate commensurate with the Larmor value (Equation [7.104]), so that

$$\left(\frac{dE'}{d\tau}\right)_{\text{rad}} = -P' \, .$$ (7.130)

Since $d\mathbf{p}' = 0$ and $d\mathbf{x}' = 0$, the transformation of the energy interval (i.e., the 0-component of p'^α) into the lab frame is simply

$$dE = \gamma \, dE' \,, \tag{7.131}$$

and similarly,

$$dt = \gamma \, d\tau \,. \tag{7.132}$$

Therefore,

$$P = P' = -\left(\frac{dE}{dt}\right)_{\text{rad}} = -\left(\frac{dE'}{d\tau}\right)_{\text{rad}} \,, \tag{7.133}$$

so that

$$\left(\frac{dE}{d\tau}\right)_{\text{rad}} = \gamma \left(\frac{dE}{dt}\right)_{\text{rad}} = -\gamma P \,. \tag{7.134}$$

Similarly, the transformation of the spatial portion of p^α here gives

$$\left(\frac{d\mathbf{p}}{d\tau}\right)_{\text{rad}} = -\frac{\gamma \vec{\beta} P}{c} \,, \tag{7.135}$$

or

$$-\left(\frac{d\mathbf{p}}{dt}\right)_{\text{rad}} = \frac{\vec{\beta}}{c} P \,. \tag{7.136}$$

Thus, in covariant form, we must have

$$\boxed{-\left(\frac{dp^\alpha}{d\tau}\right)_{\text{rad}} = \frac{P}{c^2} u^\alpha} \,. \tag{7.137}$$

The instantaneous balance of energy and momentum changes then requires that

$$\frac{dp^\alpha}{d\tau} = f^\alpha - \frac{P}{c^2} u^\alpha + \kappa^\alpha \,. \tag{7.138}$$

We already know that some 4-vector κ^α must be present in the equation of motion, because the rest mass m is known to be the same before and after the interaction and otherwise the equation would not comply with the rest-mass-preserving property $u^\alpha \, dp_\alpha/d\tau = 0$. That is,

$$u_\alpha f^\alpha - u_\alpha \frac{P}{c} u^\alpha + u_\alpha \kappa^\alpha = 0 \,, \tag{7.139}$$

so that with $u_\alpha f^\alpha = 0$, we must have

$$u_\alpha \kappa^\alpha = \frac{P}{c^2} u_\alpha u^\alpha = P . \tag{7.140}$$

In principle, Equation (7.138) is the self-consistent expression we have been looking for to describe the particle's radiative motion. It correctly takes into account the acceleration due to the external field, the reaction on the particle due to the emission of radiation, and an additional term that guarantees the observed constancy of the rest mass m. We may interpret this term as an effect resulting from the distortion of the particle's self-field, but we still need to evaluate it. Much of this rests on the nature of the Larmor power P. We know that it is a Lorentz invariant quantity, so its form written in terms of 4-vectors must be covariant. In the particle's rest frame,

$$P' = \frac{2q^2}{3c^3} |\mathbf{a}'|^2 , \tag{7.141}$$

where $\mathbf{a}'$ is its acceleration. Let us now define the 4-vector

$$a^\mu \equiv \frac{du^\mu}{d\tau} , \tag{7.142}$$

with the intention of using it to write Equation (7.141) in terms of 4D language. Evaluating a^μ in the rest frame with $v' = 0$ and $\gamma(v') = 1$, we find that $a'^\mu = (0, \mathbf{a}')$, where $\mathbf{a}'$ is the Newtonian acceleration, and so in this frame

$$P' = -\frac{2q^2}{3c^3} a'^\mu a'_\mu . \tag{7.143}$$

As we said, P is an invariant and since a^μ is a 4-vector, this form must be valid in all frames. The result of this is that we can now rewrite Equation (7.140) as

$$u_\alpha \kappa^\alpha = -\frac{2q^2}{3c^3} a^\alpha a_\alpha , \tag{7.144}$$

which becomes after integration by parts (as in the classical limit),

$$\kappa^\alpha = \frac{2q^2}{3c^3} \frac{d^2 u^\alpha}{d\tau^2} . \tag{7.145}$$

We thus arrive at the covariant Lorentz-Dirac equation

$$\boxed{m \frac{du^\alpha}{d\tau} = f^\alpha + \frac{2q^2}{3c^3} \left[\frac{d^2 u^\alpha}{d\tau^2} + \frac{u^\alpha}{c} \frac{du^\mu}{d\tau} \frac{du_\mu}{d\tau} \right]} \tag{7.146}$$

(Barut 1964; Rohrlich 1965), where f^α is given by Equation (5.133).

A slightly different way to understand our derivation is to argue that f^α cannot be the complete expression for the electromagnetic force on a charge. We know for sure that Equation (7.137) must hold true and that $dm/d\tau = 0$, i.e., the energy stored in the charge must remain untouched. Thus, the "correct" electromagnetic force f^α_{tot} must be such that $u_\alpha f^\alpha_{\text{tot}} = P u^\alpha u_\alpha / c = P \neq 0$, so that we need to make the identification

$$f^\alpha_{\text{tot}} = f^\alpha + \kappa^\alpha \,, \tag{7.147}$$

where κ^α is the missing piece of the force. In other words, we know that the charge is left unchanged after the acceleration. Therefore, the radiation field must ultimately have its origin in $F^\alpha{}_\beta$. The term κ^α represents the transition of the field from its external configuration $F^\alpha{}_\beta$ to its final radiative form.

Example 7.3. Suppose a charge q is moving relativistically in one dimension. Then, with $|\mathbf{p}| = p$, the spatial part of the Lorentz-Dirac equation becomes

$$\frac{dp}{d\tau} = \gamma F(\tau) + \frac{2q^2}{3c^3} \left[\frac{d^2}{d\tau^2}(\gamma v) + \gamma v \left(\frac{d\gamma}{d\tau} \right)^2 - \frac{\gamma v}{c^2} \left(\frac{d\gamma v}{d\tau} \right)^2 \right] \,, \tag{7.148}$$

where $F(\tau)$ is the ordinary Newtonian force as a function of τ. Thus,

$$\frac{dp}{d\tau} = \gamma F(\tau) + \frac{2q^2}{3c^3} \left[\frac{1}{m} \frac{d^2 p}{d\tau^2} + \frac{p}{m} \left(\frac{d\gamma}{d\tau} \right)^2 - \frac{p}{m^3 c^2} \left(\frac{dp}{d\tau} \right)^2 \right] \,. \tag{7.149}$$

But

$$\gamma = \sqrt{1 + \frac{p^2}{m^2 c^2}} \,, \tag{7.150}$$

so that

$$\frac{d\gamma}{d\tau} = \frac{p}{\gamma m^2 c^2} \frac{dp}{d\tau} \,. \tag{7.151}$$

It is straightforward to show that this then results in

$$\frac{p}{m} \left(\frac{d\gamma}{d\tau} \right)^2 - \frac{p}{m^3 c^2} \left(\frac{dp}{d\tau} \right)^2 = - \frac{p}{m(p^2 + m^2 c^2)} \left(\frac{dp}{d\tau} \right)^2 \,, \tag{7.152}$$

and therefore that

$$\dot{p} = \sqrt{1 + \frac{p^2}{m^2 c^2}} \, F(\tau) + \frac{2q^2}{3c^3 m} \left(\ddot{p} - \frac{p \dot{p}^2}{p^2 + m^2 c^2} \right) \,, \tag{7.153}$$

where overdots signify differentiation with respect to τ. It is common in special relativity to solve for the so-called rapidity

$$\sinh \zeta \equiv \beta \gamma \,, \tag{7.154}$$

because the corresponding equations are simpler. Let us therefore make the substitution

$$p = mc \sinh \zeta \,, \tag{7.155}$$

for which

$$\dot{p} = mc \cosh \zeta \cdot \dot{\zeta} \,, \tag{7.156}$$

and

$$\sqrt{1 + \frac{p^2}{m^2 c^2}} = \sqrt{1 + \sinh^2 \zeta} = \cosh \zeta \,. \tag{7.157}$$

Thus, combining Equations (7.153) and (7.156), and making the appropriate substitutions for p and $\dot{p}$, we get

$$m(\dot{\zeta} - \tau_c \ddot{\zeta}) = \frac{F(\tau)}{c} \,, \tag{7.158}$$

which is again the Abraham-Lorentz equation, but now for the rapidity parameter ζ instead of the velocity. The time constant is $\tau_c = 2q^2/3mc^3$.

We will in fact go ahead and solve this equation, but we can already see that there will be some limitations in its applicability, particularly near the time "boundaries." What we mean here is that normally an equation of motion is second order in time, so that two conditions (the initial position and initial velocity) are sufficient to specify the motion exactly. But now we have a third-order derivative, so that we should also prescribe the initial acceleration, which, however, depends on the net force experienced by the particle. This is a problem if, as is the customary practice, we insist on "turning" on the force at some time t_1 and then "turning" it off at a later time t_2. How do we know the acceleration at t_1 without first solving the problem? Normally, the initial conditions are completely independent of the applied force; here they are not.

The most direct way to solve Equation (7.158) is with the use of an integration factor, defined by the expression

$$\dot{\zeta} \equiv \exp(\tau/\tau_c) \, \xi \,. \tag{7.159}$$

With this substitution, Equation (7.158) becomes

$$m\dot{\xi} = -\frac{1}{c\tau_c} \exp(-\tau/\tau_c) \, F(\tau) \,, \tag{7.160}$$

which can be solved directly, and then reverting back to ζ, we get

$$m\dot{\zeta} = \frac{\exp(\tau/\tau_c)}{c\tau_c} \int_\tau^\infty \exp(-\tau''/\tau_c)\, F(\tau'')\, d\tau'' \ . \tag{7.161}$$

The upper limit on the integral was chosen with the understanding that $\dot{\zeta} \to 0$ as $\tau \to \infty$. To complete the solution, we integrate once more and get

$$\zeta(\tau) = \zeta(-\infty) + \int_{-\infty}^\tau d\tau'\ \exp(\tau'/\tau_c) \int_{\tau'}^\infty \frac{d\tau''}{\tau_c}\ \frac{F(\tau'')}{mc}\ \exp(-\tau''/\tau_c) \ . \tag{7.162}$$

With the specification of the force $F(\tau)$, the complete motion of the particle is therefore known, though one must be careful to consider the boundary problems alluded to above. For example, if we insist on letting the force be a constant F_0 over a time $0 \leq \tau \leq \tau_1$ and zero otherwise, Equations (7.156) and (7.161) result in

$$\frac{dp}{dt}(\tau < 0) = F_0 \left[1 - \exp\left(-\frac{\tau_1}{\tau_c}\right)\right] \exp\left(\frac{\tau}{\tau_c}\right) \ . \tag{7.163}$$

The implication is that the charge is "preaccelerated," since $dp/dt \neq 0$ even before the force is turned on. The appearance of τ_1 here is a direct consequence of our need to put $\dot{\zeta} = 0$ at the upper limit of integration in Equation (7.161). But suppose we really let $\tau_1 \to \infty$ so that only a dependence on τ_c survives. This solution suggests that a very rapid adjustment is taking place around the time when the force is turned on, because the particle's momentum is not being increased at the full rate until after a time $\Delta\tau \sim \tau_c$ has passed, beyond which $dp/dt \sim F_0$. Even for $v \to c$, $\Delta\tau$ is very small. For example, the distance traveled by a 100 MeV electron (with $\gamma \approx 200$) during a time $\Delta t = \gamma\tau_c$ is roughly 4×10^{-11} cm, or about 1/1000 times the nuclear diameter. Unfortunately, this makes it impractical to think about the possibility of turning the force on suddenly (i.e., within a time τ_c) at a specified time $\tau = 0$ because it would violate the uncertainty principle. Nonetheless, the character of the solution does point to this initial period when the effect of the force seems to be felt elsewhere. One interpretation is that during this time, the particle's self-field is being distorted, and radiation is subsequently generated as it attempts to recover.

8

SPECIAL TOPICS

Our development of the basic theory of electrodynamics is now complete. In this chapter, we divert our attention toward more applied aspects, including techniques to determine the angular symmetries of the field when the source distribution is extended, charged particle interactions, the dynamics of magnetized fluids, and several topics of current interest in ongoing research. Although a quantum theory of electrodynamics is essential for a better understanding of the behavior of charged particles in the extremely high energy domain, the theory in the form discussed here is highly relevant to a multitude of still-unsolved problems ranging from the behavior of high-temperature gases in confined fusion reactors to the interaction of pair plasmas with thermal and nonthermal radiation fields near the event horizon of Cygnus X-1-like black hole systems. Our intent here is to provide a flavor for how some of these topics are approached.

8.1 TIME-INDEPENDENT MULTIPOLE FIELDS

Although Equation (2.12) for the potential of an arbitrary charge distribution $\rho(\mathbf{x}')$ is exact, it tends to mask the fact that the potential and the corresponding field often possess angular symmetries linked to the charge configuration itself. In this section, we shall see how to separate out the various angular components when ρ is time independent. The much more difficult task of identifying the various multipole components in a time-dependent situation is left to the next section. Here, we seek a decomposition of the field into spherical waves with a common center, *though we must note that they are not necessarily isotropic!* Typically, the origin of our coordinate system is placed within the charge distribution, though as we shall see, the actual location of the origin is not immaterial. Starting with Equation (2.12) for the potential Φ, we look for an expansion of the distance modulus appropriate for observation distances much larger than the source size (i.e., for $|\mathbf{x}| \gg |\mathbf{x}'|$):

$$\frac{1}{|\mathbf{x} - \mathbf{x}'|} = \frac{1}{\{|\mathbf{x}|^2 + |\mathbf{x}'|^2 - 2\mathbf{x} \cdot \mathbf{x}'\}^{1/2}}$$

$$= \frac{1}{|\mathbf{x}|} \left\{ 1 + \frac{|\mathbf{x}'|^2}{|\mathbf{x}|^2} - \frac{2\mathbf{x} \cdot \mathbf{x}'}{|\mathbf{x}|^2} \right\}^{-1/2} . \tag{8.1}$$

For large distances, the expansion parameter $|\mathbf{x}'|/|\mathbf{x}|$ is much smaller than 1, and so by the binomial theorem

$$\frac{1}{|\mathbf{x} - \mathbf{x}'|} \approx \frac{1}{|\mathbf{x}|} \left\{ 1 + \frac{\mathbf{x} \cdot \mathbf{x}'}{|\mathbf{x}|^2} - \frac{1}{2} \frac{|\mathbf{x}'|^2}{|\mathbf{x}|^2} + \frac{3}{2} \frac{(\mathbf{x} \cdot \mathbf{x}')^2}{|\mathbf{x}|^4} \right\} . \tag{8.2}$$

The dependence of each of these terms on the source position $\mathbf{x}'$ is now very simple, and it is straightforward to carry out the integration in Equation (2.12):

$$\Phi(\mathbf{x}) = \frac{q}{r} + \sum_i \frac{x_i p_i}{r^3} + \frac{1}{2} \sum_{i,j} Q_{ij} \frac{x_i x_j}{r^5} + \dots , \tag{8.3}$$

where

$$q \equiv \int \rho(\mathbf{x}') \, d^3 x' , \tag{8.4}$$

$$p_i = \int x_i' \, \rho(\mathbf{x}') \, d^3 x' , \tag{8.5}$$

and

$$Q_{ij} = \int (3x_i' x_j' - r'^2 \, \delta_{ij}) \, \rho(\mathbf{x}') \, d^3 x' , \tag{8.6}$$

are, respectively, the monopole moment q, the dipole moment $\mathbf{p}$, and the quadrupole moment tensor Q_{ij}. As written, Equation (8.3) is a multipole expansion of the potential written in rectangular coordinates. Its usefulness is apparent the moment we realize that only one or maybe two terms dominate the potential for typical source configurations. As an example, consider a source $\rho(\mathbf{x}') = q\delta^3(\mathbf{x}')$, for which

$$\Phi(\mathbf{x}) = \frac{q}{r} + 0 + 0 + \dots , \tag{8.7}$$

i.e., only the term describing the (radial) spherically symmetric contribution to Φ survives. However, when $\rho(\mathbf{x}') = q\delta^3(\mathbf{x}_1) - q\delta^3(\mathbf{x}_2)$, then

$$\Phi(\mathbf{x}) = 0 + \frac{\mathbf{x} \cdot \mathbf{p}}{r^3} + 0 + \ldots, \tag{8.8}$$

where now

$$\mathbf{p} \equiv q\mathbf{x}_1 - q\mathbf{x}_2 . \tag{8.9}$$

In these two simple examples, the angular dependence of the field is rather obvious. The field is independent of angle when the source is a monopole, and it has a straightforward $\cos\theta$ dependence when the charge is separated out, e.g., as collinear points, θ being the angle between the observation point vector $\mathbf{x}$ and the line passing through the charges.

In extreme cases, the angular dependence of Φ may be more complicated, and in such instances, it is common to expand the potential using spherical harmonics:

$$\Phi(\mathbf{x}) = \sum_{l=0}^{\infty} \sum_{m=-l}^{l} q_{lm} \frac{Y_{lm}(\theta, \phi)}{r^{l+1}} \frac{4\pi}{2l+1} , \tag{8.10}$$

where q_{lm} now constitute the multipole moments. Using the expansion

$$\frac{1}{|\mathbf{x} - \mathbf{x}'|} = 4\pi \sum_{l=0}^{\infty} \sum_{m=-l}^{l} \frac{1}{2l+1} \frac{r_<^l}{r_>^{l+1}} Y_{lm}^*(\theta', \phi') Y_{lm}(\theta, \phi) \tag{8.11}$$

in Equation (2.12), where $r_<$ $(r_>)$ is the smaller (larger) of $|\mathbf{x}|$ and $|\mathbf{x}'|$, we compare the two expressions for Φ term by term, and thus infer that

$$q_{lm} = \int Y_{lm}^*(\theta', \phi') \, r'^l \, \rho(\mathbf{x}') \, d^3x' , \tag{8.12}$$

which are coefficients that clearly contain information about the θ-dependence through l and the ϕ-dependence through m. Note that the number of "multipole" moments does not depend on which coordinate system we use. For example, the number of dipole moments p_i (i.e., three) is the same as the number of q_{1m}, and the number of components of the quadrupole tensor Q_{ij} (i.e., five independent components, since this is a trace-free symmetric tensor) is the same as the number of q_{2m}.

The most direct way of recovering the multipole field is to differentiate the appropriate term in the expansion for Φ:

$$\mathbf{E} = -\vec{\nabla}\Phi , \tag{8.13}$$

where

$$E_r(l,m) = \frac{4\pi(l+1)}{2l+1} \, q_{lm} \, \frac{Y_{lm}(\theta,\phi)}{r^{l+2}} \; , \tag{8.14}$$

$$E_\theta(l,m) = -\frac{4\pi}{2l+1} \, q_{lm} \, \frac{1}{r^{l+2}} \, \frac{\partial Y_{lm}(\theta,\phi)}{\partial\theta} \; , \tag{8.15}$$

$$E_\phi(l,m) = -\frac{4\pi}{2l+1} \, q_{lm} \, \frac{1}{r^{l+2}} \, \frac{im}{\sin\theta} \, Y_{lm}(\theta,\phi) \; . \tag{8.16}$$

Having said this, we must also recognize that the multipole moments do depend on the choice of origin. This is not surprising since the angular distribution of the source itself depends on where the origin is placed, and we recognize the fact that the field symmetries ultimately depend on the configuration of the charge. Looking at the clump in Figure 8.1, for example, it is obvious that the angular description of $\rho(\mathbf{x}')$ depends on whether the reference point is at $\vec{0}$ or $\vec{0}'$. As a concrete example, let us see how a displacement of the origin by a constant vector $\vec{\Delta}'$ alters a given moment, in this case, the dipole moment $\mathbf{p}$ given in Equation (8.5).

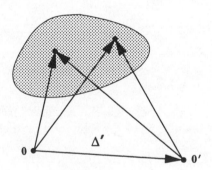

Figure 8.1   Calculation of the multipole moments of an arbitrary charge distribution with respect to two different origins.

Starting with the general expression as applied to $\vec{0}$, we have

$$\begin{aligned}
\mathbf{p}_0 &= \int \mathbf{x}\,\rho(\mathbf{x})\,d^3x \\[2mm]
&= \int (\mathbf{x}' + \vec{\Delta}')\rho(\mathbf{x}')\,d^3x' \; .
\end{aligned} \tag{8.17}$$

That is,

$$\mathbf{p}_0 = \int \mathbf{x}' \rho(\mathbf{x}') \, d^3x' + \vec{\Delta}' \int \rho(\mathbf{x}') \, d^3x' \, , \tag{8.18}$$

where $\vec{\Delta}'$ is assumed to be constant. Thus, using the definition of $\mathbf{p}$, we see that

$$\mathbf{p}_0 = \mathbf{p}_{0'} + \vec{\Delta}' \cdot q \, , \tag{8.19}$$

which clearly depends on $\vec{\Delta}'$. However, notice that the moment q is not changed when the origin is moved. This is an example of a general theorem, which says that the lowest nonvanishing multipole moment of any charge distribution is independent of the choice of origin, but that all higher moments are in general dependent on where the coordinate system is centered. Thus, the practical use of multipole expansions is really limited to situations where the charge density exhibits some degree of symmetry, so that its "center" is well defined.

8.2   MULTIPOLE EXPANSION OF TIME-DEPENDENT FIELDS

Let us now extend these ideas to the case where the fields are no longer static. In the free space between sources, the four components of the potential A^μ must satisfy the wave equations (3.13) and (3.14). If we decompose the potential into a series of time-harmonic terms (i.e., a Fourier series), each of these frequency components

$$A_\omega^\mu(\mathbf{x}, t) = A_{0\omega}^\mu(\mathbf{x}) \exp(-i\omega t) \tag{8.20}$$

is then a solution of the Helmholtz equation

$$(\vec{\nabla}^2 + k^2) A_{0\omega}^\mu(\mathbf{x}) = 0 \, , \tag{8.21}$$

where as always,

$$k^2 = \frac{\omega^2}{c^2} \, . \tag{8.22}$$

The customary method of solution is to separate out the angular and radial variables with the expansion

$$A_{0\omega}^\mu(\mathbf{x}) = \sum_{lm} R_l(r) \, Y_{lm}(\theta, \phi) \ \text{(for each } \mu) \, , \tag{8.23}$$

which results in two equations:

$$-\left(\frac{1}{\sin\theta} \frac{\partial}{\partial\theta} \sin\theta \frac{\partial}{\partial\theta} + \frac{1}{\sin^2\theta} \frac{\partial^2}{\partial\phi^2} \right) Y_{lm} = l(l+1) Y_{lm} \tag{8.24}$$

(the angular equation), and

$$\left[-\frac{1}{r}\frac{d^2}{dr^2}r + \frac{l(l+1)}{r^2} \right] R_l = k^2 R_l(r) \tag{8.25}$$

(the radial-wave equation). Very importantly, all the l-solutions to Equation (8.25) are independent of m, and so the radial-wave factors will be the same for all the $2l+1$ differently oriented angular patterns. This is a property that time-dependent multipole waves share with the static multipole fields we considered earlier.

Like the Bessel equation, the radial-wave equation has solutions that depend on r only through the product $\zeta \equiv kr$, since this substitution in (8.25) gives

$$-\left[\frac{d^2}{d\zeta^2} - \frac{l(l+1)}{\zeta^2} \right](\zeta R_l) = (\zeta R_l) . \tag{8.26}$$

This means that radial waves differ from each other only in the *scale* of their extensions into space as measured by r in units of their wavelength $\lambda = 2\pi/k$, but not their profiles. In the simplest case, $l = 0$, the solutions are trivially found to be

$$\zeta R_0 \sim \exp(\pm i\zeta) , \tag{8.27}$$

describing radial waves that propagate isotropically, since $Y_{00} = 1/\sqrt{4\pi}$ is independent of direction. The $l = 0$ solutions are thus the spherical Hankel functions of order zero:

$$h_0(\zeta) = \frac{\exp(i\zeta)}{i\zeta} , \tag{8.28}$$

and the most general *isotropic* ($l = 0$) solution is therefore

$$R_0 = a_0 h_0 + b_0 h_0^* , \tag{8.29}$$

where a_0 and b_0 are coefficients to be determined by the boundary conditions for each value of μ. As it turns out, this vacuum spherically symmetric solution is unphysical (i.e., it cannot be generated in a manner that is self-consistent with the actual boundary conditions) since it is not possible to vary the charge of a point source in time while still preserving overall charge conservation. As such, the only real solution to Maxwell's equations with spherical symmetry is the time-independent Coulomb solution.

For general values of l, the solution to the radial-wave equation can be written

$$R_l(r) = a_l\, h_l(kr) + b_l\, h_l^*(kr) \text{ (for each } \mu) , \tag{8.30}$$

where $h_l(\zeta)$ are the spherical Hankel functions of order l. This R_l is expressed as an arbitrary superposition of outgoing and incoming waves.

Let us now return to the angular equation (8.24), which is often written as

$$L^2 Y_{lm} = l(l+1)Y_{lm} \,, \tag{8.31}$$

where

$$L^2 \equiv -\left[\frac{1}{\sin\theta}\frac{\partial}{\partial\theta}\left(\sin\theta\frac{\partial}{\partial\theta}\right) + \frac{1}{\sin^2\theta}\frac{\partial^2}{\partial\phi^2}\right] \,. \tag{8.32}$$

When expressed in the form

$$\mathbf{L} \equiv \frac{1}{i}\left(\mathbf{r} \times \vec{\nabla}\right) \,, \tag{8.33}$$

it is apparent that $\mathbf{L}$ is proportional to the orbital angular-momentum operator in quantum mechanics. We will find this expression useful in our derivation of the fields below.

In a source-free region of empty space, time-harmonic fields satisfy the equations

$$\vec{\nabla} \times \mathbf{E} = ik\,\mathbf{B} \,,$$

$$\vec{\nabla} \cdot \mathbf{E} = 0\,\mathbf{B} \,,$$

$$\vec{\nabla} \times \mathbf{B} = -ik\,\mathbf{E} \,,$$

$$\vec{\nabla} \cdot \mathbf{B} = 0 \,. \tag{8.34}$$

We can either eliminate $\mathbf{E}$, so that

$$(\vec{\nabla}^2 + k^2)\mathbf{B} = 0 \,,$$

$$\vec{\nabla} \cdot \mathbf{B} = 0 \,,$$

$$\mathbf{E} = \frac{i}{k}\vec{\nabla} \times \mathbf{B} \,, \tag{8.35}$$

or we can eliminate $\mathbf{B}$ and get

$$(\vec{\nabla}^2 + k^2)\mathbf{E} = 0 \,,$$

$$\vec{\nabla} \cdot \mathbf{E} = 0 \,,$$

$$\mathbf{B} = -\frac{i}{k}\vec{\nabla} \times \mathbf{E} \,. \tag{8.36}$$

In either case, each of the Cartesian components of $\mathbf{E}$ and $\mathbf{B}$ satisfies the Helmholtz equation, which is solved by the techniques we have just been describing. The solutions must also automatically satisfy the conditions $\vec{\nabla} \cdot \mathbf{E} = \vec{\nabla} \cdot \mathbf{B} = 0$. An elegant way to handle this is to define the scalar quantities $\mathbf{r} \cdot \mathbf{E}$ and $\mathbf{r} \cdot \mathbf{B}$, which have the property that

$$\vec{\nabla}^2(\mathbf{r} \cdot \mathbf{E}) = \mathbf{r} \cdot (\vec{\nabla}^2 \mathbf{E}) + 2\vec{\nabla} \cdot \mathbf{E} , \tag{8.37}$$

and similarly for $\mathbf{B}$.[21] The Helmholtz equation for $\mathbf{r} \cdot \mathbf{E}$ is therefore

$$(\vec{\nabla}^2 + k^2)(\mathbf{r} \cdot \mathbf{E}) = \mathbf{r} \cdot (\vec{\nabla}^2 \mathbf{E}) + 2\vec{\nabla} \cdot \mathbf{E} + k^2 \mathbf{r} \cdot \mathbf{E} . \tag{8.38}$$

But from Equation (8.36) we know that $\vec{\nabla}^2 \mathbf{E} = -k^2 \mathbf{E}$, and so as long as $\mathbf{r} \cdot \mathbf{E}$ is a solution to the Helmholtz equation, the zero divergence condition on $\mathbf{E}$ is satisfied self-consistently. The same applies for $\mathbf{B}$, and this result also ensures that all the components have the same l and m.

Our experience with the potentials allows us to immediately write down the solutions for $\mathbf{r} \cdot \mathbf{E}$ and $\mathbf{r} \cdot \mathbf{B}$, including a summation over all possible values of the expansion indices:

$$\mathbf{r} \cdot \mathbf{E} = \sum_{l,m} [a_{lm}\, h_l(kr) + b_{lm}\, h_l^*(kr)]\, Y_{lm}(\theta, \phi) , \tag{8.39}$$

$$\mathbf{r} \cdot \mathbf{B} = \sum_{l',m'} [a_{l'm'}\, h_{l'}(kr) + b_{l'm'}\, h_{l'}^*(kr)]\, Y_{l'm'}(\theta, \phi) , \tag{8.40}$$

where each $[a_{lm}\, h_l(kr) + b_{lm}\, h_l^*(kr)]\, Y_{lm}(\theta, \phi)$ (and similarly for the quantities with primed indices) must represent a multipole field of order (l, m). A "pure" magnetic multipole field of order (l, m) is usually defined by the condition

$$\mathbf{r} \cdot \mathbf{E}_{lm}^{(M)} = 0 , \tag{8.41}$$

for which the corresponding magnetic field is then

$$\mathbf{r} \cdot \mathbf{B}_{lm}^{(M)} \equiv \frac{l(l+1)}{k}\, g_{lm}(kr)\, Y_{lm}(\theta, \phi) , \tag{8.42}$$

and

$$g_{lm}(kr) \equiv c_{lm}\, h_l(kr) + d_{lm}\, h_l^*(kr) . \tag{8.43}$$

However, this is not yet sufficient information for us to determine the fields $\mathbf{E}$ and $\mathbf{B}$ themselves. For this, we must introduce the additional constraints

---

[21] For a review of several different, but equivalent, approaches to follow in this derivation, see Bouwkamp and Casimir (1954).

(8.36) imposed by Maxwell's equations. In particular, using the last expression in (8.36), we can say that

$$kr \cdot \mathbf{B} = \frac{1}{i}\mathbf{r} \cdot (\vec{\nabla} \times \mathbf{E}) = \frac{1}{i}(\mathbf{r} \times \vec{\nabla}) \cdot \mathbf{E} = \mathbf{L} \cdot \mathbf{E} \ . \tag{8.44}$$

The last step follows from the same type of vector manipulation used in Equation (2.112). Thus, in order for us to have a "pure" magnetic multipole as we have here defined it, we need the electric field to satisfy not only Equation (8.41), but also

$$\mathbf{L} \cdot \mathbf{E}_{lm}^{(M)} = l(l+1)\, g_{lm}(kr)\, Y_{lm}(\theta,\phi) \ . \tag{8.45}$$

Does it make sense to have a nonzero electric field when all we're trying to set up is a magnetic multipole field? Well, think about the situation we're dealing with here. The source is varying in time, and we expect the fields to be time dependent as well. In Chapter 3 we learned that time-varying fields act as sources for each other, and so it would be inconsistent for us to have a time-dependent **B** multipole and no associated electric field at all. It should not be surprising, therefore, that we end up with a description that looks like a cross between the electrostatic multipole fields and the transverse propagating fields in a wave guide (see § 4.5). Since **L** operates only on Y_{lm}, the solution to Equation (8.45) is

$$\mathbf{E}_{lm}^{(M)} = g_{lm}(kr)\, \mathbf{L}\, Y_{lm}(\theta,\phi) \tag{8.46}$$

because $\mathbf{L}^2 Y_{lm} = l(l+1)Y_{lm}$. The corresponding **B** field is then

$$\mathbf{B}_{lm}^{(M)} = -\frac{i}{k}\vec{\nabla} \times \mathbf{E}_{lm}^{(M)} \ , \tag{8.47}$$

and the coefficients in Equation (8.43) should be matched accordingly to be consistent with this.

Clearly, the same derivation applies to "pure" electric multipole fields, yielding the field configuration

$$\mathbf{B}_{lm}^{(E)} = f_{lm}(kr)\, \mathbf{L}\, Y_{lm}(\theta,\phi) \ , \tag{8.48}$$

and

$$\mathbf{E}_{lm}^{(E)} = \frac{i}{k}\vec{\nabla} \times \mathbf{B}_{lm}^{(E)} \ , \tag{8.49}$$

where $f_{lm}(kr)$ is analogous to $g_{lm}(kr)$.

The most general solution we can write down for a time-dependent multipole electromagnetic field is therefore a superposition over all such components of order (l, m), encompassing both "pure" electric and "pure" magnetic multipoles:

$$\mathbf{E} = \sum_{l,m} \left[\frac{i}{k} a_{lm}^E \vec{\nabla} \times f_l(kr) \, \mathbf{L} \, Y_{lm} + a_{lm}^M g_l(kr) \, \mathbf{L} \, Y_{lm} \right] , \qquad (8.50)$$

$$\mathbf{B} = \sum_{l,m} \left[a_{lm}^E f_l(kr) \, \mathbf{L} \, Y_{lm} - \frac{i}{k} a_{lm}^M \vec{\nabla} \times g_l(kr) \, \mathbf{L} \, Y_{lm} \right] , \qquad (8.51)$$

and this now becomes a boundary-value problem which requires evaluation of the coefficients a_{lm}^E and a_{lm}^M. Notice that we have now denoted the functions f and g only by the index l, since the m-dependence is contained solely within the Y_{lm} and the constants a_{lm}^E and a_{lm}^M take into account the different weighting factors for the various spherical harmonics (see Blatt and Weisskopf 1952; Morse and Feshbach 1953).

Example 8.1. Consider a harmonically time-dependent electric field $\mathbf{E}$, which is known to have the functional form

$$\mathbf{E} = E_0 \sin \theta \, \hat{\theta} \qquad (8.52)$$

at $r = R$ near the source. Let us use the formalism we have developed in this section to determine the multipole field structure in the *radiation* zone. For simplicity, we shall assume that there is no current or magnetization.

We begin with the general solutions given in Equations (8.50) and (8.51). In the radiation zone, the fields depend on the imposed boundary conditions at smaller radii. Since the source is localized to a region $R_0 \ll r$, let's assume that we have only outgoing waves. In that case,

$$f_l(kr) \sim h_l(kr) . \qquad (8.53)$$

But in the radiation zone (i.e., $kr \gg 1$),

$$h_l(kr) \to (-i)^{l+1} \frac{\exp(ikr)}{kr} . \qquad (8.54)$$

In addition, we have no current or magnetization, so that only electric multipole fields should be present, i.e., $a_{lm}^M = 0$. Thus,

$$\mathbf{B} \to \mathbf{B}_E = \sum_{l.m} a_{lm}^E (-i)^{l+1} \frac{\exp(ikr)}{kr} \, \mathbf{L} \, Y_{lm} . \qquad (8.55)$$

The corresponding electric field is

$$\mathbf{E} \to \mathbf{E}_E = \sum_{l,m} a_{lm}^E \frac{(-i)^l}{k^2} \left[\vec{\nabla} \left(\frac{\exp(ikr)}{r} \right) \times \mathbf{L} \, Y_{lm} + \frac{\exp(ikr)}{r} \vec{\nabla} \times \mathbf{L} \, Y_{lm} \right] .$$

(8.56)

But

$$\vec{\nabla} \equiv \hat{r} \frac{\partial}{\partial r} - \frac{i}{r^2} \mathbf{r} \times \mathbf{L} ,$$

(8.57)

so that

$$i \vec{\nabla} \times \mathbf{L} = \mathbf{r} \, \vec{\nabla}^2 - \vec{\nabla} \left(1 + r \frac{\partial}{\partial r} \right) .$$

(8.58)

And so, keeping only terms of order $1/r$,

$$\mathbf{E}_E \sim - \sum_{l,m} (-i)^{l+1} \frac{\exp(ikr)}{kr} \left[\hat{n} \times \mathbf{L} Y_{lm} - \frac{1}{k} \left(\mathbf{r} \vec{\nabla}^2 - \vec{\nabla} \right) Y_{lm} \right] a_{lm}^E , \quad (8.59)$$

where $\hat{n} \equiv \mathbf{r}/r$. Since the second term is of order $1/kr$ times the first, we have in the limit $kr \gg 1$,

$$\mathbf{E}_E \to - \sum_{lm} (-i)^{l+1} \frac{\exp(ikr)}{kr} \hat{n} \times \mathbf{L} Y_{lm} \, a_{lm}^E$$

$$= -\hat{n} \times \mathbf{B}_E = \mathbf{B}_E \times \hat{n} .$$

(8.60)

These are clearly typical radiation fields, transverse to the radius vector, and falling off as $1/r$.

To solve for the fields, let's take $\hat{n} \times$ Equation (8.59), for then

$$\hat{n} \times \mathbf{E}_E = - \sum_{l,m} (-i)^{l+1} \frac{\exp(ikr)}{kr} a_{lm}^E \{ (\hat{n} \cdot \mathbf{L}) \hat{n} - (\hat{n} \cdot \hat{n}) \mathbf{L} \} Y_{lm}$$

$$= \sum_{l,m} (-i)^{l+1} \frac{\exp(ikr)}{kr} a_{lm}^E \, \mathbf{L} Y_{lm} ,$$

(8.61)

since $\hat{n} \cdot \mathbf{L} = \mathbf{r} \cdot (\mathbf{r} \times \vec{\nabla})/i = \vec{\nabla} \cdot (\mathbf{r} \times \mathbf{r})/i = 0$. Therefore,

$$a_{lm}^E (-i)^{l+1} \frac{\exp(ikr)}{kr} = \frac{1}{\sqrt{l(l+1)}} \int Y_{lm}^* \mathbf{L} \cdot (\hat{n} \times \mathbf{E}_E) \, d\Omega ,$$

(8.62)

where

$$\mathbf{L} \cdot (\hat{n} \times \mathbf{E}_E) = \frac{1}{i} (\mathbf{r} \times \vec{\nabla}) \cdot (\hat{n} \times \mathbf{E}_E) .$$

(8.63)

The fact that we know what $\mathbf{E}_E$ is at some given radius $r = R$ thus allows us to evaluate the expansion coefficients outright. Putting

$$\mathbf{r} \times \mathbf{E}_E|_R = RE_0 \sin\theta \, \hat{\phi} \tag{8.64}$$

and

$$
\begin{aligned}
(\mathbf{r} \times \vec{\nabla}) \cdot (\hat{n} \times \mathbf{E}_E) &= \left(\hat{\phi}\frac{\partial}{\partial\theta} - \hat{\theta}\frac{1}{\sin\theta}\frac{\partial}{\partial\phi} \right) \cdot \left(E_0 \sin\theta \, \hat{\phi} \right) \\
&= E_0 \cos\theta \\
&= \sqrt{\frac{4\pi}{3}} \, E_0 \, Y_{10} \ ,
\end{aligned}
\tag{8.65}
$$

right away we get

$$a_{lm}^E = \frac{1}{(-i)^{l+1}} \frac{kR}{\exp(ikR)} \frac{-i}{\sqrt{l(l+1)}} \int Y_{lm} \sqrt{\frac{4\pi}{3}} E_0 \, Y_{10} \, d\Omega \ . \tag{8.66}$$

Thus,

$$a_{10}^E = \frac{iE_0 kR}{\exp(ikR)} \sqrt{\frac{2\pi}{3}} \ , \tag{8.67}$$

and all others are zero. The complete solution is

$$\mathbf{B}_E = -iE_0 \sqrt{\frac{2\pi}{3}} \left(\frac{R}{r} \right) \exp\{ik(r - R)\} \mathbf{L} \, Y_{10} \ , \tag{8.68}$$

and

$$\mathbf{E}_E = \mathbf{B}_E \times \hat{n} \ . \tag{8.69}$$

8.3   COLLISIONS BETWEEN CHARGED PARTICLES

In Chapter 7, we examined the bremsstrahlung emissivity of an electron interacting electromagnetically with an ion. The radiation produced by this form of acceleration is just one of the physical manifestations resulting from the "collision" between charges. Another very important aspect of this problem is to determine the energy exchange between swiftly moving, charged particles and the deflections from their incident direction. Consider the particular example of two charges q and Q, with respective masses m and M, moving initially with velocities **v** and **V** in the laboratory frame as shown in Figure 8.2. As we did in § 7.4, it will be convenient for us to transform to another frame in which one of the particles is at rest. Taking this to be m's frame (see Figure 8.3), M here moves with a velocity $\mathbf{V}'$, where

$$\mathbf{V}'_{\parallel} = \frac{\mathbf{V}_{\parallel} - \mathbf{v}}{1 - \mathbf{V} \cdot \mathbf{v}/c^2} \, , \tag{8.70}$$

$$\mathbf{V}'_{\perp} = \frac{\mathbf{V}_{\perp}}{\gamma(v)(1 - \mathbf{V} \cdot \mathbf{v}/c^2)} \, , \tag{8.71}$$

and $\gamma(v) \equiv (1 - v^2/c^2)^{-1/2}$.

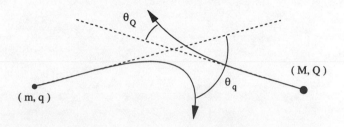

Figure 8.2 Collision between two charged particles with mass m and M, causing deflections by angles θ_q and θ_Q, respectively.

For definiteness, let us assume that **V** and **v** are antiparallel (along the x' axis) and in the $x' - z'$ plane, and that the particle paths are initially separated by a distance b. Then,

$$V'_{\perp} = 0 \, , \tag{8.72}$$

and

$$V'_{\parallel} = \frac{V + v}{1 + vV/c^2} \, . \tag{8.73}$$

Since b is perpendicular to **v**, this separation remains the same in the primed (i.e., m's) frame.

Fortunately, we have already gone through the process of finding the transformed fields due to particle Q in q's rest frame (see Equations [7.77]), so using the charge Q instead of Ze as we did earlier, we know that q sees the field components

$$E'_{Qx} = \frac{Q\gamma(V')\,V'\,t'}{[b^2 + \gamma^2(V')\,V'^2\,t'^2]^{3/2}}\,,$$

$$E'_{Qz} = \frac{Q\gamma(V')\,b}{[b^2 + \gamma^2(V')\,V'^2\,t'^2]^{3/2}}\,,$$

$$E'_{Qy} = 0\,. \tag{8.74}$$

It should be stressed here that this derivation is valid only as long as Q moves with minimal changes to its velocity. These expressions give the electric field components in q's initial frame of reference, in terms of Q's initial velocity. Thus, if during the interaction Q experiences a substantial change to its velocity, the transformed field changes at each step of the interaction. We therefore require that $M \gg m$, as was the case for the ion and electron system that led to the derivation of Equation (7.74).

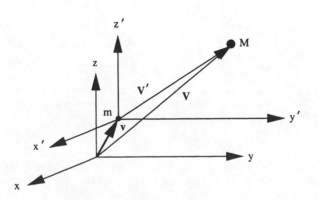

Figure 8.3 Transformation of the coordinates from the laboratory (un-primed) frame to m's rest (primed) frame. In this frame, moving with velocity **v** relative to the laboratory, M has a velocity **V**′.

The change in momentum experienced by q is

$$\Delta \mathbf{p}'_q = \int_{-\infty}^{\infty} q\,\mathbf{E}'_Q \cdot dt'\,. \tag{8.75}$$

This expression itself is only valid as long as $\Delta \mathbf{p}'_q$ is sufficiently small that $\mathbf{v}' \times \mathbf{B}'$ forces in q's initial rest frame may be ignored during the interaction. Since E'_x is antisymmetric in t', we don't have to evaluate the integral to realize that $\Delta p'_{qx} = 0$. Only the transverse component survives:

$$\Delta p'_{qz} = \int_{-\infty}^{\infty} \frac{qQ\gamma(V')\,b}{[b^2 + \gamma^2(V')\,V'^2\,t'^2]^{3/2}}\,dt'$$

$$= \frac{qQb}{V'} \int_{-\infty}^{\infty} \frac{du}{(b^2 + u^2)^{3/2}} = \frac{2qQ}{bV'}\,. \tag{8.76}$$

Notice that most of the frame dependence has dropped out. In particular, it seems at first remarkable that $\Delta p'$ does not depend on $\gamma(V')$. This is a consequence of the fact that although E' is "bunched up" (near $t' = 0$) and enhanced by a factor $\gamma(V')$, the duration of the pulse is decreased by the same factor due to time dilation. The resultant effect (which is the time-integrated force) is therefore independent of $\gamma(V')$.

Thus, as a result of the interaction, the energy of q is changed to the new value

$$U'_q = \left\{ m^2 c^4 + \left(\frac{2qQ}{bV'} \right)^2 c^2 \right\}^{1/2}\,. \tag{8.77}$$

As long as $\Delta p'$ is small, a binomial expansion simplifies this to

$$U'_q = mc^2 + \frac{2q^2Q^2}{mV'} \left(\frac{1}{b^2} \right)\,. \tag{8.78}$$

Thus, in the laboratory (i.e., unprimed) frame after the collision,

$$U_q/c = \gamma(v) \left(\frac{U'_q}{c} + \frac{\mathbf{v} \cdot \mathbf{p}'}{c} \right)\,,$$

$$\mathbf{P}_{\parallel} = \gamma(v) \left(\mathbf{p}'_{\parallel} + \frac{v}{c}\frac{U'_q}{c} \right)\,,$$

$$\mathbf{P}_{\perp} = \mathbf{P}'_{\perp}\,, \tag{8.79}$$

where $\mathbf{p}' = (0, 0, 2qQ/bV')$. Remembering that $\mathbf{v}$ was chosen to be antiparallel to $\mathbf{V}$ and that consequently $\mathbf{v} \cdot \mathbf{p}' = 0$, we see that

$$\frac{U_q}{c} = \gamma(v)\frac{U'_q}{c}\,,$$

$$\mathbf{P}_{\parallel} = \gamma(v)\frac{\mathbf{v}}{c}\frac{U'_q}{c} = \frac{\mathbf{v}}{c}\frac{U_q}{c}\,,$$

$$\mathbf{P}_{\perp} = \left(0, 0, \frac{2qQ}{bV'} \right)\,. \tag{8.80}$$

Particle m's altered energy in the laboratory frame is therefore seen to be

$$U_q = \gamma(v)\, mc^2 + \frac{2q^2Q^2}{mV'^2}\left(\frac{1}{b^2}\right)\gamma(v)\,, \tag{8.81}$$

and using the transformation of velocity (8.73), we therefore infer an interaction energy exchange

$$\Delta U_q = \frac{2q^2Q^2}{mb^2}\frac{(1 + vV/c^2)^2}{(V + v)^2}\,\gamma(v)\,. \tag{8.82}$$

The deflection angle θ_q is known from the change in vector momentum of particle m. Thus,

$$\tan\theta_q = \frac{|\mathbf{p}_\perp|}{|\mathbf{p}_\parallel|}$$

$$\approx \left[\frac{2qQ}{b}\frac{1 + vV/c^2}{V + v}\right][\gamma(v)\, m\, v]^{-1}\,. \tag{8.83}$$

When both v and V are relativistic, then

$$\Delta U_q \approx \frac{2q^2Q^2\gamma(v)}{mb^2c^2}\,, \tag{8.84}$$

and

$$\tan\theta_q \approx \frac{2qQ}{\gamma(v)\, bmc^2}\,. \tag{8.85}$$

Both ΔU_q and θ_q decrease with b, as expected. In addition, θ_q decreases with swiftness, whereas ΔU_q increases with $\gamma(v)$ because of the effects of boosting. The caveat is that our analysis breaks down for small values of b, where ΔU_q becomes larger than the available energy.

8.4   MAGNETOHYDRODYNAMICS

Interesting new physics is encountered when highly conducting fluid media interact with magnetic fields. Much of our attention in the early chapters was devoted to the effects on charged particles due to the dynamical influence of external fields, and correspondingly, we learned a great deal about the fields produced by charges. In Chapter 7 we started to examine the self-consistent dynamical behavior of radiating charges, taking into account the influence of momentum and energy losses to the radiation. Magnetohydrodynamics is the study of the dynamics of magnetized (though highly conducting) fluids in which the charge densities and currents produce macroscopic fields and are affected by them. This self-consistent coupling of the bulk charge motions and the associated fields is in some sense a synthesis of some of these ideas, but in order to keep the problem tractable, we shall consider only the nonrelativistic aspects of the interaction for which the radiative reaction effects are negligible. Still, this will constitute our first attempt at using the Maxwell and dynamics equations in a consistent manner, since we here permit the charges and currents to interact with their own fields.

Consider the Faraday field equation

$$\vec{\nabla} \times \mathbf{E} = -\frac{1}{c}\frac{\partial \mathbf{B}}{\partial t} , \tag{8.86}$$

and suppose that our medium is moving with some velocity **v** with respect to the laboratory. In the local rest frame of the medium, Ohm's law says

$$\mathbf{J}' = \sigma\,\mathbf{E}' , \tag{8.87}$$

where prime denotes rest frame coordinates. From the Lorentz transformations (Equation [5.131]) we know that

$$\mathbf{E}' = \gamma\left(\mathbf{E} + \frac{\mathbf{v}}{c}\times\mathbf{B}\right) , \tag{8.88}$$

so that for nonrelativistic motion,

$$\mathbf{J}' \approx \sigma\left(\mathbf{E} + \frac{\mathbf{v}}{c}\times\mathbf{B}\right) . \tag{8.89}$$

Thus, if the medium is neutral so that there are no advected currents,

$$\mathbf{J} \approx \sigma\left(\mathbf{E} + \frac{\mathbf{v}}{c}\times\mathbf{B}\right) . \tag{8.90}$$

Equations (8.86) and (8.90) may be combined to eliminate one of the unknowns, usually taken to be the electric field **E**. The result is

$$\frac{\partial \mathbf{B}}{\partial t} = \vec{\nabla}\times\left(\mathbf{v}\times\mathbf{B} - \frac{c}{\sigma}\mathbf{J}\right) . \tag{8.91}$$

It must be emphasized again that the source appearing in this equation is itself subject to the effects of the field. One way to put this in explicitly is to use another of Maxwell's equations (Ampère's law) in the form

$$\vec{\nabla}\times\mathbf{B} = \frac{4\pi}{c}\mathbf{J} . \tag{8.92}$$

The displacement current is usually ignored because $\partial \mathbf{E}/\partial t$ is $O(v^2/c^2)$ compared to $c\vec{\nabla} \times \mathbf{B}$. Then,

$$\frac{\partial \mathbf{B}}{\partial t} = \vec{\nabla} \times (\mathbf{v} \times \mathbf{B}) - \frac{c^2}{4\pi\sigma}(\vec{\nabla} \times \vec{\nabla} \times \mathbf{B})$$

$$= \vec{\nabla} \times (\mathbf{v} \times \mathbf{B}) - \frac{c^2}{4\pi\sigma}\{\vec{\nabla}(\vec{\nabla} \cdot \mathbf{B}) - \vec{\nabla}^2\mathbf{B}\} . \tag{8.93}$$

Thus, since $\vec{\nabla} \cdot \mathbf{B} = 0$,

$$\boxed{\frac{\partial \mathbf{B}}{\partial t} = \vec{\nabla} \times (\mathbf{v} \times \mathbf{B}) + \frac{c^2}{4\pi\sigma}\vec{\nabla}^2\mathbf{B}} . \tag{8.94}$$

This differential equation is a complete description of the magnetic field within the fluid in terms of its velocity and conductivity. Below, we consider the behavior of $\mathbf{B}$ in two important limiting situations.

Special Case 1. We here consider an environment with very high conductivity. When $\sigma \to \infty$, clearly

$$\frac{\partial \mathbf{B}}{\partial t} \approx \vec{\nabla} \times (\mathbf{v} \times \mathbf{B}) . \tag{8.95}$$

Note that although we derived this using Ohm's law (to get Equation [8.94]), its validity requires only that the fluid cannot support any significant electric field in its own reference frame, i.e., that $\mathbf{E}' \approx 0$ in Equation (8.88). The result in (8.95) then follows directly from Faraday's law (Equation [8.86]). To see what this means physically, let us consider the magnetic flux through a loop moving with the medium[22] (see Figure 8.4). Then,

$$\frac{d}{dt}\int_C \mathbf{B} \cdot d\mathbf{A} = \int_C \left(\frac{\partial}{\partial t} + \mathbf{v} \cdot \vec{\nabla}\right) \mathbf{B} \cdot d\mathbf{A} , \tag{8.96}$$

where $d/dt \equiv \partial/\partial t + \mathbf{v} \cdot \vec{\nabla}$ is the convective derivative. Suppose we now consider a small volume of fluid within which $\mathbf{v}$ is only slowly varying. In that case, spatial derivatives of $\mathbf{v}$ can be made arbitrarily small compared to the other derivatives (e.g., $\partial \mathbf{B}/\partial t$), and therefore

$$\mathbf{v} \cdot \vec{\nabla}\mathbf{B} \approx \vec{\nabla} \times (\mathbf{B} \times \mathbf{v}) + \mathbf{v}(\vec{\nabla} \cdot \mathbf{B}) = \vec{\nabla} \times (\mathbf{B} \times \mathbf{v}) . \tag{8.97}$$

---

[22]A very readable and detailed development of these ideas is given in the excellent book by Parker (1979).

Thus,

$$\frac{d}{dt} \int_C \mathbf{B} \cdot d\mathbf{A} = \int_C \left[\frac{\partial \mathbf{B}}{\partial t} + \vec{\nabla} \times (\mathbf{B} \times \mathbf{v}) \right] \cdot d\mathbf{A} = 0 . \qquad (8.98)$$

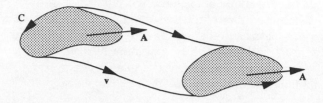

Figure 8.4 Transport of a magnetic flux loop C with cross-sectional area $\mathbf{A}$ through a conducting medium. The loop's velocity is $\mathbf{v}$.

That is, when the conductivity σ is very large, the magnetic flux through any loop moving with the medium is constant in time. The way to visualize this is to think of the lines of force as being *frozen* into the fluid and being carried along with it. Because $\sigma \to \infty$, any motion of the magnetic field through the fluid results in the generation of $\mathbf{v} \times \mathbf{B}$ "eddy" currents that act to maintain the status quo, and because of the high degree of conductivity, it costs virtually nothing energetically for this to happen.

Special Case 2. Let us now consider the other extreme—a system with very high resistivity. When the conductivity σ is very small,

$$\frac{\partial \mathbf{B}}{\partial t} \approx \frac{c^2}{4\pi\sigma} \vec{\nabla}^2 \mathbf{B} , \qquad (8.99)$$

which is a *diffusion* equation. In this case, the magnetic field is not "slaved" to the material motion, but it instead diffuses (or decays) away on a time scale τ_D, where

$$\frac{|\mathbf{B}|}{\tau_D} \approx \frac{c^2}{4\pi\sigma} \frac{|\mathbf{B}|}{L^2} , \qquad (8.100)$$

and L is a length characteristic of the spatial variation of $\mathbf{B}$. That is,

$$\boxed{\tau_D \approx 4\pi\sigma \, L^2/c^2} . \qquad (8.101)$$

For example, a 1 cm copper sphere has a diffusion time scale $\tau_D \approx 1$ second. By comparison, the sun's magnetic field is expected to diffuse away in about 10^{10} years.

The behavior of the fluid-field system is determined primarily by which of these regimes we are in. The study of magnetohydrodynamics is quite extensive and continues to be an active area of research, particularly in the context of confined high-temperature plasmas, both here on earth and in the strong gravitational environments near black holes. The magnetic field in these systems is quite strong and can significantly influence the gas dynamics via its pressure. We know that in a fluid with gas pressure P, the force density on it can be written $-\vec{\nabla}P$. A similar construction is possible for the magnetic field. To see how this comes about, we recognize that the electromagnetic field exerts a force density

$$\frac{1}{c}(\mathbf{J} \times \mathbf{B}) = -\frac{1}{4\pi}\mathbf{B} \times (\vec{\nabla} \times \mathbf{B}) \qquad (8.102)$$

(from Ampère's law). Thus, the electromagnetic force density can itself be written, via the identity

$$\vec{\nabla}(\mathbf{B} \cdot \mathbf{B}) = 2(\mathbf{B} \cdot \vec{\nabla})\mathbf{B} + 2\mathbf{B} \times (\vec{\nabla} \times \mathbf{B}) , \qquad (8.103)$$

as

$$\frac{1}{c}(\mathbf{J} \times \mathbf{B}) = -\vec{\nabla}\left(\frac{B^2}{8\pi}\right) + \frac{1}{4\pi}\left(\mathbf{B} \cdot \vec{\nabla}\right)\mathbf{B} . \qquad (8.104)$$

We therefore associate a magnetic pressure

$$\boxed{P_M \equiv B^2/8\pi} \qquad (8.105)$$

with $\mathbf{B}$, and in cases where $(\mathbf{B} \cdot \vec{\nabla})\mathbf{B} = 0$ (see the discussion following Equation [8.106] below), the magneto-gas *equilibrium* condition is achieved when

$$P + P_M = \text{constant} . \qquad (8.106)$$

The second term on the right-hand side of Equation (8.104) also has a direct physical interpretation. If we write $\mathbf{B} \cdot \vec{\nabla}\mathbf{B}/4\pi$ in the form $(|\mathbf{B}|\partial|\mathbf{B}|/\partial s)/4\pi$, where ds is an element of path length along the direction of $\mathbf{B}$, then it can be further transformed into $(\partial|\mathbf{B}|^2/\partial s)/8\pi$. That is, this term is the spatial rate of change of magnetic field energy density along $\hat{s}$, which therefore represents a tensile force in that direction.

8.5 ALFVÉN WAVES AND PARTICLE ACCELERATION

Quasars are believed to be prodigious powerhouses energized by massive black holes surrounded by disks of infalling matter. The fact that we see gamma-rays with energies $\epsilon_\gamma \gtrsim 30$ GeV coming from their direction suggests that the particles producing these photons have a Lorentz factor $\gamma \sim 30$ GeV $/mc^2$, which for electrons is $\gtrsim 6 \times 10^4$. Other evidence, such as the superluminal motion observed in many of these sources (see § 8.7 below), also implies that γ should be much larger than 1. The Lorentz factor may be even higher in pulsars—rapidly spinning, magnetized neutron stars—reaching values beyond $\sim 10^7$ or more. It is clear, therefore, that nongravitational acceleration schemes must play a role in energizing the particles in many of these systems. Perhaps the most common of these is electromagnetic acceleration.

A key ingredient in this subject is the role played by the magnetic field. Several different mechanisms may contribute to the acceleration, depending on the field distribution, since charges generally move in circles perpendicular to **B** because of the Lorentz $\mathbf{v} \times \mathbf{B}$ force, as well as moving more freely along **B**. When **B** is random, the principal method of acceleration is the Fermi process, in which bundles of magnetic flux act as mirrors bouncing the particles back and forth and thereby energizing them. On the other hand, when the field is well organized, a more direct acceleration mechanism is based on the idea that electric fields may be generated parallel to **B**, where the particle motion is unrestricted. This is the situation we have here, and to understand how a magnetic field disturbance energizes the charges, let us for the moment return to some of the equations we developed in the previous section on magnetohydrodynamics. When the medium is highly conducting, Equation (8.89) shows that

$$\mathbf{E} \approx -\frac{\mathbf{v}}{c} \times \mathbf{B} . \tag{8.107}$$

Thus, from Faraday's law (8.86), the simple equation satisfied by the magnetic field in these circumstances is

$$\frac{\partial \mathbf{B}}{\partial t} = \vec{\nabla} \times (\mathbf{v} \times \mathbf{B}) . \tag{8.108}$$

At the same time, the gas is subject to the mass conservation law,

$$\frac{\partial \eta}{\partial t} + \vec{\nabla} \cdot (\eta \, \mathbf{v}) = 0 , \tag{8.109}$$

and the dynamical influence of the electric and magnetic fields. (The mass density η is not to be confused with the charge density ρ.) That is, starting from the Lorentz force density (which is a continuum generalization of the point particle Equation [1.3])

$$\eta \frac{d\mathbf{v}}{dt} = \rho\, \mathbf{E} + \frac{1}{c}(\mathbf{J} \times \mathbf{B})\,, \qquad (8.110)$$

we assume that the medium is neutral (i.e., that the charge density ρ is zero) and take $\mathbf{J}$ to be given by Ampère's law (Equation [1.34]), with the result that

$$\eta \frac{\partial \mathbf{v}}{\partial t} + \eta(\mathbf{v} \cdot \vec{\nabla})\mathbf{v} = -\frac{1}{4\pi}\,\mathbf{B} \times \left[\vec{\nabla} \times \mathbf{B} - \frac{1}{c}\frac{\partial \mathbf{E}}{\partial t}\right]\,. \qquad (8.111)$$

As we shall see, the acceleration scheme associated with this magnetic field disturbance acts on a small fraction of charges within the plasma (i.e., those particles moving principally in the direction along $\mathbf{B}$). Most of the plasma responds as a fluid with a subrelativistic velocity. Equation (8.111) describes the behavior of this slowly moving gas, and so for a nonrelativistic motion, we reduce this equation to

$$\eta \frac{\partial \mathbf{v}}{\partial t} + \eta(\mathbf{v} \cdot \vec{\nabla})\mathbf{v} \approx -\frac{1}{4\pi}\,\mathbf{B} \times [\vec{\nabla} \times \mathbf{B}]\,. \qquad (8.112)$$

Equations (8.108), (8.109), and (8.112) form a complete set of coupled relations describing the behavior of the fluid and the magnetic field in an interacting environment.

Suppose now that the field $\mathbf{B}$ is jiggled, perhaps due to a disturbance at $z = 0$ as shown in Figure 8.5. This field might, for example, be anchored in the disk surrounding massive, compact objects or might be frozen into the surface layers of highly magnetized neutron stars, as discussed above. The solution to Equations (8.108), (8.109), and (8.112) that describes this wave propagation with perfect symmetry in the $x - y$ plane is

$$\mathbf{B} = B_0 \hat{z} + B_{\mathrm{A}}\, \exp(ikz - i\omega t)\, \hat{x}\,, \qquad (8.113)$$

$$\mathbf{v} = 0 + v_{\mathrm{A}}\, \exp(ikz - i\omega t)\, \hat{x}\,, \qquad (8.114)$$

$$\mathbf{E} = 0 + E_{\mathrm{A}}\, \exp(ikz - i\omega t)\, \hat{y}\,, \qquad (8.115)$$

where clearly B_{A}, v_{A}, and E_{A} represent the traveling perturbation. It is not difficult to see with direct substitution of (8.113)–(8.115) into the coupled equations that

$$\mathbf{B_{\mathrm{A}}} \equiv B_{\mathrm{A}}\, \exp(ikz - i\omega t)\, \hat{x} = -\frac{k}{\omega}\, B_0\, \mathbf{v}\,. \qquad (8.116)$$

Not surprisingly, this is a plane wave—a so-called Alfvén wave—with phase velocity $v_\alpha = \omega/k$ traveling in the z-direction. But notice that $\mathbf{E}$ is always perpendicular to $\mathbf{B}$ and hence can only accelerate charges perpendicular to the magnetic field. This does not lead to persistent acceleration because of $\mathbf{v} \times \mathbf{B}$ forces.

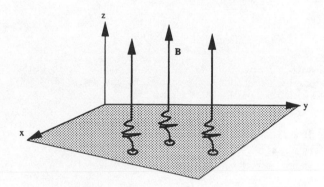

Figure 8.5 A strong, uniform magnetic field $\mathbf{B}$ anchored in a highly conducting medium is jiggled at the base. Like a perturbation on a string, this disturbance travels outward along the field lines.

Real situations, however, are hardly sufficiently ideal to sustain perfect plane waves. Instead, some $x - y$ structure is expected, and because curl $\mathbf{B}$ is then nonzero, we expect a current and therefore an electric field component along $\hat{z}$. The MHD equations ([8.108], [8.109], and [8.112]) also allow for the following solution:

$$\mathbf{B} = B_0\hat{z} + B_\mathrm{A} \sin(k_y y) \exp(ikz - i\omega t)\, \hat{x} , \qquad (8.117)$$

which we may interpret as one of the frequency components in a Fourier series expansion for a more general structure of the field in the y-direction. Then, from Equation (8.86),

$$\frac{\partial \mathbf{E}}{\partial t} \approx c\,\frac{\partial B_x}{\partial z}\,\hat{y} - c\,\frac{\partial B_x}{\partial y}\,\hat{z} , \qquad (8.118)$$

which for a harmonic field leads to

$$\mathbf{E} \approx \frac{ic}{\omega}\,[ikB_\mathrm{A}\,\sin(k_y y)\,\hat{y} - k_y B_\mathrm{A}\,\cos(k_y y)\,\hat{z}]\,\exp(ikz - i\omega t) . \qquad (8.119)$$

The important thing to notice is the appearance of an E_z component, which can accelerate particles along the (local) $\hat{z}$ direction (see Figure 8.6).

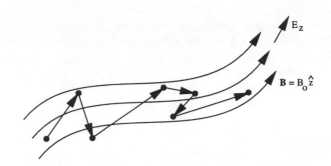

Figure 8.6 The presence of an electric field component E_z in the direction of **B** induces an acceleration of the charges, though restricted by a drag due to collisions with other particles in the medium.

The collision frequency ν_c in a typical neutron-star magnetosphere, where the particle density is $\sim 10^{16} - 10^{26}$ cm^{-3}, is about 10^7 s^{-1}. In between scatterings, the E_z component accelerates the charges according to

$$\frac{dp_z}{dt} = q\,E_z \,, \tag{8.120}$$

where $p_z = \gamma m\,v_z$. Thus, for a relativistic electron (with $v_z \approx c =$ constant),

$$\frac{d\gamma}{dt} \approx \frac{e\,E_z}{m_e\,c} \,. \tag{8.121}$$

As a rough estimate of how energetic the particles can become, we take these relations and put

$$\gamma_{\mathrm{max}} \sim \frac{eE_z}{m_e c\nu_c} \approx \frac{ek_y B_{\mathrm{A}}}{m_e \omega \nu_c} \,. \tag{8.122}$$

In typical pulsars, $k_y \sim 2\pi/\lambda_{\mathrm{crust}}$, where the crust scale length λ_{crust} is of order 10 cm, and B_{A} ranges from 10^8 G to as high as the underlying magnetic field strength $\sim 10^{12}$ G. In addition, $\omega \approx v_\alpha k$, where $k \sim 2\pi/R_*$ and the stellar radius is $R_* \approx 10$ km. In principle, this mechanism can therefore accelerate particles to a Lorentz factor much in excess of 10^{10}, since the Alfvén wave *phase* velocity v_α is greater than c. In practice, several damping influences and the creation of lepton pairs set in well before this plateau is reached.

8.6   SYNCHROTRON EMISSION

In Chapter 7, we considered the radiation produced by electrons accelerated in the Coulomb field of an ion, a process known as bremsstrahlung. A similarly common phenomenon is the acceleration of charges in a magnetic field, which leads to another form of emissivity called cyclotron radiation, when the particle motion is nonrelativistic, and synchrotron radiation in the fully relativistic limit (Schott 1912; Rybicki and Lightman 1979). This mechanism is sometimes also referred to as "magnetic" bremsstrahlung, by analogy with the process we considered earlier. The composite radiation spectrum depends quite strongly on the energy partitioning among the particles, and in nature we encounter both thermal (usually Maxwellian) and nonthermal (usually power-law) distributions. In this section, we shall derive the single particle emissivity and spectrum. Once this is known, it is a straightforward matter to integrate this expression over the particle energy occupation function to obtain the plasma's overall radiation profile for any given physical conditions.

Let us begin by considering the behavior of a nonrelativistic electron gyrating in a magnetic field, as shown in Figure 8.7.

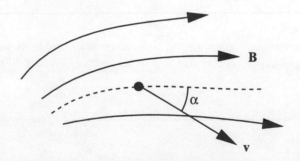

Figure 8.7 A nonrelativistic electron gyrating with velocity **v** in a magnetic field **B**. The angle between **v** and the local field lines is α.

From the Lorentz force law (Equation [1.3]) we know that the electron experiences a force

$$\mathbf{F}_B = (-e) \left(\frac{\mathbf{v}}{c} \times \mathbf{B} \right) \, , \tag{8.123}$$

which points in a direction perpendicular to both **v** and **B**. As such, the motion of the electron is a superposition of a translational path with (constant) velocity $v_\| = v \cos \alpha$, together with a circular motion with velocity $v_\perp = v \sin \alpha$. The electron is constantly being accelerated, and it therefore emits radiation whose energy source is ultimately the agent that energizes the particle. The equation of motion gives

$$m_e \mathbf{a} = \frac{-e}{c} \mathbf{v} \times \mathbf{B} \, , \tag{8.124}$$

or

$$\frac{m_e\,v_\perp^2}{r_{\text{gyr}}} = \left|\frac{-ev_\perp B}{c}\right| = \left|\frac{-e\,v\,\sin\alpha\,B}{c}\right| . \tag{8.125}$$

The particles therefore gyrate with a radius

$$r_{\text{gyr}} = \frac{v\,\sin\alpha}{\omega_{\text{gyr}}} , \tag{8.126}$$

where the gyration frequency is

$$\omega_{\text{gyr}} \equiv \frac{eB}{m_e c} = 1.8 \times 10^7 \left(\frac{B}{1\,\text{gauss}}\right) \quad \text{radians s}^{-1} . \tag{8.127}$$

The emitted power for a single accelerated charge is given by the Larmor rate (Equation [7.104]). Thus, with an acceleration $a = \omega_{\text{gyr}}\,v\,\sin\alpha$, an individual charge emits at a rate

$$P = \frac{2e^2}{3c^3}\,\omega_{\text{gyr}}^2\,v^2\,\sin^2\alpha . \tag{8.128}$$

This produces the well-known cyclotron radiation, which is monochromatic with a frequency $\omega = \omega_{\text{gyr}}$ due to the perfect circular motion of the electron in the $v_\parallel$ frame (Figure 8.8).

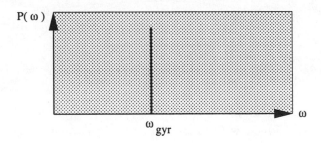

Figure 8.8 The cyclotron spectrum of an electron moving nonrelativistically.

In addition, the emission is fairly isotropic, arising from the dipole pattern affixed to the acceleration vector, as shown in Figure 8.9(a).

An arguably more interesting situation arises when the particle motion is relativistic, for then the particle's trajectory is distorted by time-dilation and length-contraction effects. The particle no longer exhibits perfect circular motion and so will emit at more than one frequency. The components of the particle's equation of motion in terms of its charge $q = -e$ are now

$$\frac{d}{dt}(\gamma m_e \mathbf{v}) = \frac{q}{c} \mathbf{v} \times \mathbf{B} , \tag{8.129}$$

and

$$\frac{d}{dt}(\gamma m_e c^2) = q \mathbf{v} \cdot \mathbf{E} = 0 . \tag{8.130}$$

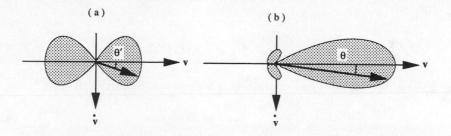

Figure 8.9 Angular distribution of the radiation (a) in the electron's instantaneous rest (i.e., primed) frame, and (b) in the laboratory frame, where it is moving with velocity **v**. The angle θ' is that measured in the electron's rest frame, whereas θ is that seen in the laboratory frame.

The second of these forces the condition $\gamma = $ constant, so that in the first equation,

$$m_e \gamma \frac{d\mathbf{v}}{dt} = \frac{q}{c} \mathbf{v} \times \mathbf{B} . \tag{8.131}$$

That is,

$$\frac{d\mathbf{v}_\parallel}{dt} = 0 , \tag{8.132}$$

and

$$\frac{d\mathbf{v}_\perp}{dt} = \frac{q}{\gamma m_e c} \mathbf{v}_\perp \times \mathbf{B} , \tag{8.133}$$

where $\mathbf{v}_\parallel$ is parallel to **B** and $\mathbf{v}_\perp$ is the perpendicular component. It is evident that the electron's frequency of gyration is now

$$\omega_B = \left| \frac{qB}{\gamma m_e c} \right| , \tag{8.134}$$

and that its acceleration is

$$a_\perp = \omega_B \, v_\perp .$$ (8.135)

Thus,

$$
P = \frac{2e^2}{3c^3} \, \gamma^4 \, \frac{e^2 B^2}{\gamma^2 \, m_e^2 c^2} \, v_\perp^2
$$

$$
= \frac{2}{3} \, r_0^2 \, c \beta_\perp^2 \, \gamma^2 \, B^2 ,
$$ (8.136)

where $r_0 \equiv e^2/m_e c^2$ is the classical electron radius. It is necessary to average this expression over all angles for a given normalized speed β. Using the pitch angle α shown in Figure 8.7, we have

$$
\langle \beta_\perp^2 \rangle = \frac{\beta^2}{4\pi} \int \sin^2 \alpha \, d\Omega = \frac{2\beta^2}{3} ,
$$ (8.137)

and therefore the average Larmor power from a single electron is

$$
\langle P \rangle = \frac{4}{3} \sigma_T \, c \, \beta^2 \, \gamma^2 \, u_B ,
$$ (8.138)

where $u_B \equiv B^2/8\pi$ is the magnetic field energy density and

$$
\sigma_T \equiv \frac{8\pi \, r_0^2}{3}
$$ (8.139)

is the Thomson scattering cross section.

Getting the single particle *spectrum* of this radiation is equally important, but a little more involved. Before actually carrying out the derivation, we can foresee the direction in which this will take us by noting the importance of beaming effects, primarily due to aberration resulting from the relativistic transformation of angles. As we have already mentioned earlier, the angular distribution of radiation in the electron's rest frame is the dipolar pattern centered on $\dot{\mathbf{v}}$, as shown in Figure 8.9(a). In terms of the angle θ' defined in this figure, the angular distribution of the radiation in the electron's (primed) frame is

$$
S' \propto \sin^2 \left(\frac{\pi}{2} - \theta' \right) = \cos^2 \theta'
$$ (8.140)

(see Equation [4.176]). But this angle changes for an observer in the laboratory frame, and so the power distribution is shifted. The easiest way to understand how this comes about is to think about how a pulse of light is observed in the two frames. We recall that one of the postulates of special relativity is that c

is the same for all observers, in all frames. Suppose now we have a pulse of light traveling in a straight line making an angle θ' with respect to the frame velocity $\mathbf{v}$. Its velocity components parallel and perpendicular to $\mathbf{v} \equiv v\hat{z}$ are clearly

$$c'_z = c\cos\theta' \, , \tag{8.141}$$

and

$$c'_\perp = c\sin\theta' \, . \tag{8.142}$$

According to Equations (5.38), (5.39), and (5.40), the photon's velocity components in the laboratory frame are therefore

$$c_z = \frac{c'_z + v}{\gamma(1 + vc'_z/c^2)} \, , \tag{8.143}$$

and

$$c_\perp = \frac{c'_\perp}{\gamma(1 + vc'_z/c^2)} \, . \tag{8.144}$$

But $c_\perp = c\sin\theta$ in the laboratory frame, and so we arrive at the angle transformation formula

$$\sin\theta = \frac{\sin\theta'}{\gamma(1 + \beta\cos\theta')} \, . \tag{8.145}$$

Returning now to Equation (8.140), we see that the flux drops to half its maximum value in the electron's rest frame at $\theta' = \pi/4$. But according to (8.145), this point corresponds to $\sin\theta \approx \theta \approx 1/\gamma$ in the laboratory frame, and so we anticipate that most of the emitted radiation is beamed into a cone of angle $\sim 1/\gamma$ about $\mathbf{v}$, as shown in Figure 8.9(b). The main result of this is that whereas in the nonrelativistic case the observed electric field is sinusoidal in time (producing the monochromatic power spectrum shown in Figure 8.8), here the $1/\gamma$ beaming leads to a spike in the observed field intensity that requires many frequency components to adequately describe its shape. Since, in addition, the gyration frequency $\omega_{\mathrm{gyr}} \to \omega_B = \omega_{\mathrm{gyr}}/\gamma$, the contributing frequencies are much closer together (as we shall see by the end of this section).

Our formal procedure for deriving the single-particle synchrotron spectrum begins with expression (7.88) for the energy observed in the laboratory frame per unit area per unit angular frequency ω, written in terms of the Fourier transform of the electric field. Making the additional change from area to solid angle like we did in the derivation of Equation (7.64), we have

$$\frac{dW}{d\omega \, d\Omega} = \frac{c}{4\pi^2} \left| \int_{-\infty}^{\infty} [R^2 \mathbf{E}(t)]_{\tilde{t}} \, \exp(i\omega t) \, dt \right|^2$$

$$= \frac{q^2}{4\pi^2 c} \left| \int_{-\infty}^{\infty} \left[\frac{\hat{n} \times \left\{ (\hat{n} - \vec{\beta}) \times \dot{\vec{\beta}} \right\}}{(1 - \hat{n} \cdot \vec{\beta})^3} \right]_{\tilde{t}} \exp(i\omega t) \, dt \right|^2 , \qquad (8.146)$$

where the expression in square brackets is to be evaluated at the retarded time $\tilde{t} = t - R(\tilde{t})/c$ (see § 7.3). Thus, changing the integration variable to $\tilde{t}$, putting $dt = (1 - \hat{n} \cdot \vec{\beta}) d\tilde{t}$, and noting that $R(t') \approx |\mathbf{r}| - \hat{n} \cdot \mathbf{r}_q(\tilde{t})$ far from the source at $\mathbf{r}_q(\tilde{t})$, we have

$$\frac{dW}{d\omega \, d\Omega} = \frac{1}{4\pi^2 c} \left| \int_{-\infty}^{\infty} \frac{\hat{n} \times \left\{ (\hat{n} - \vec{\beta}) \times \dot{\vec{\beta}} \right\}}{(1 - \hat{n} \cdot \vec{\beta})^2} \right.$$

$$\left. \times \exp[i\omega(\tilde{t} - \hat{n} \cdot \mathbf{r}_q(\tilde{t})/c)] \, d\tilde{t} \right|^2 . \qquad (8.147)$$

This can be simplified somewhat by noting that $\hat{n} \times \{ (\hat{n} - \vec{\beta}) \times \dot{\vec{\beta}} \}/(1 - \hat{n} \cdot \vec{\beta})^2 = d/d\tilde{t}[\hat{n} \times (\hat{n} \times \vec{\beta})]$, for then integrating (8.147) by parts gives

$$\frac{dW}{d\omega \, d\Omega} = \frac{q^2 \omega^2}{4\pi^2 c} \left| \int_{-\infty}^{\infty} \hat{n} \times (\hat{n} \times \vec{\beta}) \, \exp[i\omega(\tilde{t} - \hat{n} \cdot \mathbf{r}_q(\tilde{t})/c)] \, d\tilde{t} \right|^2 . \qquad (8.148)$$

Putting $\hat{\epsilon}_\parallel \equiv \hat{n} \times \hat{\epsilon}_\perp$, we see from Figure 8.10 that

$$\hat{n} \times (\hat{n} \times \vec{\beta}) = -\hat{\epsilon}_\perp \sin\left(\frac{v\tilde{t}}{r_B} \right) + \hat{\epsilon}_\parallel \cos\left(\frac{v\tilde{t}}{r_B} \right) \sin\theta , \qquad (8.149)$$

where $\beta \approx 1$ has been used. For the argument of the exponential, we note that

$$\tilde{t} - \frac{\hat{n} \cdot \mathbf{r}_q(\tilde{t})}{c} = \tilde{t} - \frac{r_B}{c} \cos\theta \sin\left(\frac{v\tilde{t}}{r_B} \right)$$

$$\approx \frac{1}{2\gamma^2} \left\{ (1 + \gamma^2\theta^2)\tilde{t} + \frac{c^2\gamma^2\tilde{t}^3}{3r_B^2} \right\} , \qquad (8.150)$$

where the sine and cosine functions have been expanded for small arguments and it has been assumed that $v \approx c$. Expanding the sine and cosine functions once more in Equation (8.149) and returning to (8.148), we find that the radiation spectrum may be written as the sum of contributions from the two polarization states

$$\frac{dW_\perp}{d\omega\, d\Omega} = \frac{q^2\omega^2}{4\pi^2 c} \left| \int_{-\infty}^{\infty} \frac{c\tilde{t}}{r_B} \exp\left\{ \frac{i\omega}{2\gamma^2} \left[(1+\gamma^2\theta^2)\tilde{t} + \frac{c^2\gamma^2\tilde{t}^3}{3r_B^2} \right] \right\} d\tilde{t} \right|^2 , \quad (8.151)$$

and

$$\frac{dW_\parallel}{d\omega\, d\Omega} = \frac{q^2\omega^2\theta^2}{4\pi^2 c} \left| \int_{-\infty}^{\infty} \exp\left\{ \frac{i\omega}{2\gamma^2} \left[(1+\gamma^2\theta^2)\tilde{t} + \frac{c^2\gamma^2\tilde{t}^3}{3r_B^2} \right] \right\} d\tilde{t} \right|^2 , \quad (8.152)$$

with

$$\frac{dW}{d\omega\, d\Omega} = \frac{dW_\perp}{d\omega\, d\Omega} + \frac{dW_\parallel}{d\omega\, d\Omega} . \quad (8.153)$$

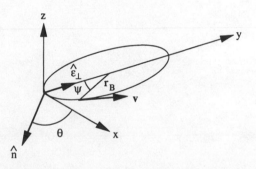

Figure 8.10 Synchrotron radiation from a particle passing the point $\mathbf{x} = 0$ at the retarded time $\tilde{t} = 0$. Here, r_B is the radius of curvature, $\mathbf{v}$ is the particle's velocity, $\psi = v\tilde{t}/r_B$, and $\hat{\epsilon}_\perp$ is a polarization vector perpendicular to $\mathbf{v}$ and in the orbital plane at $\tilde{t} = 0$ (see § 4.2).

These expressions for the spectral components can be greatly simplified with the change of variable

$$x \equiv \frac{\gamma c\tilde{t}}{r_B\,(1 + \gamma^2\theta^2)^{1/2}} . \quad (8.154)$$

In addition, we know that most of the radiation will be produced in the "forward" (i.e., $\theta \approx 0$) direction (see Figure 8.9b), so that

$$\frac{\omega r_B(1 + \gamma^2\theta^2)^{3/2}}{3c\gamma^3} \approx \frac{\omega}{2\omega_c} , \quad (8.155)$$

where the critical angular frequency is

$$\omega_c \equiv \frac{3}{2}\gamma^3\omega_B \sin\alpha \, , \tag{8.156}$$

and α is the pitch angle, as shown in Figure 8.7. In writing Equation (8.155), we have made use of the definition

$$r_B = \frac{v\sin\alpha}{\omega_B} \, , \tag{8.157}$$

which is the relativistic generalization of (8.126) with $\omega_{\text{gyr}} \to \omega_B = \omega_{\text{gyr}}/\gamma$. Thus, the spectral components may be written

$$\frac{dW_\perp}{d\omega\,d\Omega} = \frac{q^2\omega^2}{3\pi^2 c}\left[\frac{r_B(1+\gamma^2\theta^2)}{\gamma^2 c}\right]^2 K_{2/3}^2(\omega/2\omega_c) \, , \tag{8.158}$$

and

$$\frac{dW_\parallel}{d\omega\,d\Omega} = \frac{q^2\omega^2\theta^2}{3\pi^2 c}\left[\frac{r_B(1+\gamma^2\theta^2)^{1/2}}{\gamma c}\right]^2 K_{1/3}^2(\omega/2\omega_c) \, , \tag{8.159}$$

where

$$K_{2/3}(\eta) \equiv \int_{-\infty}^{\infty} x\,\exp\left[\frac{3}{2}i\eta\left(x+\frac{1}{3}x^3\right)\right]\,dx \tag{8.160}$$

and

$$K_{1/3}(\eta) \equiv \int_{-\infty}^{\infty} \exp\left[\frac{3}{2}i\eta\left(x+\frac{1}{3}x^3\right)\right]\,dx \tag{8.161}$$

are the modified Bessel functions of order 2/3 and 1/3, respectively. In the final step, these spectral components must be integrated over all solid angles Ω to give the energy per angular frequency range emitted by the particle per complete orbit:

$$\frac{dW_\perp}{d\omega} = \frac{2q^2\omega^2 r_B^2 \sin\alpha}{3\pi c^3\gamma^4}\int_{-\infty}^{\infty}(1+\gamma^2\theta^2)^2 K_{2/3}^2(\omega/2\omega_c)\,d\theta \, , \tag{8.162}$$

$$\frac{dW_\parallel}{d\omega} = \frac{2q^2\omega^2 r_B^2 \sin\alpha}{3\pi c^3\gamma^2}\int_{-\infty}^{\infty}(1+\gamma^2\theta^2)\theta^2 K_{1/3}^2(\omega/2\omega_c)\,d\theta \, . \tag{8.163}$$

The result is

$$\frac{dW_\perp}{d\omega} = \frac{\sqrt{3}\,q^2\gamma\sin\alpha}{2c}\left[F(\omega/2\omega_c)+G(\omega/2\omega_c)\right] \, , \tag{8.164}$$

$$\frac{dW_\parallel}{d\omega} = \frac{\sqrt{3}\,q^2\gamma\sin\alpha}{2c}\left[F(\omega/2\omega_c)-G(\omega/2\omega_c)\right] \, , \tag{8.165}$$

where

$$F(\omega/2\omega_c) \equiv \left(\frac{\omega}{\omega_c}\right) \int_{\omega/\omega_c}^{\infty} K_{5/3}(\xi)\, d\xi \,, \tag{8.166}$$

and

$$G(\omega/2\omega_c) \equiv \left(\frac{\omega}{\omega_c}\right) K_{2/3}\left(\frac{\omega}{\omega_c}\right) \,. \tag{8.167}$$

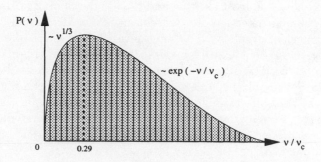

Figure 8.11 The single-particle spectrum in the limit of very high velocity. The spectrum involves a great number of harmonics, which blend together to form a continuous envelope due to the action of several broadening mechanisms. One such mechanism occurs for a distribution of particle energies, for which the spectra of particles do not all fall on the same lines. Here, $\nu = \omega/2\pi$, $\nu_c = \omega_c/2\pi$.

The expressions we have derived give the total energy emitted per angular frequency per complete particle orbit. In order to convert these to the power per angular frequency, we divide them by the orbital period $T = 2\pi/\omega_B$ and add the two polarization components to reach the total emitted power

$$P(\omega) = \frac{\sqrt{3}q^3 B \sin\alpha}{2\pi m_e c^2} F(\omega/2\omega_c) \,. \tag{8.168}$$

This function is shown in Figure 8.11, together with its limiting forms in the very low and very high frequency regions. The maximum occurs at $\omega = 0.29\omega_c$.

8.7 ECHOES OF THE BIG BANG

The ideas we have introduced and developed in this book are applicable to many areas in physics, such as the one described here and those discussed in the next two sections. One of the most exciting applications—certainly the most intriguing—is to our ongoing study of the early universe. Astronomy is slowly enlarging the windows available for observing the cosmos with the introduction of instruments that detect particles other than photons. For example, the air shower arrays that measure the debris of collisions between cosmic rays and particles in the upper atmosphere are gaining in sensitivity and spatial resolution and are now providing important information on high-energy particle acceleration and transport in the interstellar medium. Currently, neutrino astronomy is one of the most exciting and rapidly developing fields, and we are likely to see the first detection of extrasolar neutrinos (from steady sources, as opposed to transient events such as supernova 1987A) within the next decade or two. However, no one can deny the fact that most of the information we acquire from the cosmos is borne by the radiation in transit to the earth from distant objects. The oldest source about which we know is the Big Bang. How these photons propagate through the universe, how their polarization changes when they pass through a magnetized region, and how they interact with matter, can tell us a great deal about the fabric of spacetime and the genesis of everything around us.

There are very few "fossils" that we can use to unravel the mystery of the early universe; one of the most important is the cosmic microwave background radiation (CMBR). In the theory of the Big Bang, the early universe was hot and dense, and most of the hydrogen was ionized. So the free electrons and protons formed a hot sea of charges and currents coupled to electric and magnetic fields, forming an equilibrium with the photon distribution via interactions that include Compton scattering, and bremsstrahlung (see § 7.4) and synchrotron (see § 8.6) emission and absorption processes. In the beginning, matter and radiation were coupled strongly and behaved as a single fluid in thermodynamic equilibrium, so the radiation field was that of a blackbody.

The rapid expansion that ensued lowered the matter density and temperature, and about a month after the Big Bang the photon creation and annihilation rates would no longer have been fast enough to guarantee a blackbody spectrum, so the lack of distortions in the CMBR places severe limits on any processes that would have added or subtracted energy from the radiation field. (An excellent, though somewhat technical, account of the early history of the universe may be found in Linde 1990.) The very existence of the CMBR (discovered serendipitously by Penzias and Wilson in 1965 and measured to high precision with the *COBE* satellite by Fixsen et al. in 1996) appears to rule out the steady state model of the universe, for which changes occur without a gradient in time, so that any one step in the evolution of the cosmos would look like any other. This picture is inconsistent with the CMBR since the universe today does not contain the isothermal and opaque environment required to produce a blackbody spectrum.

Instead, several arguments favor an inflationary cosmology, in which the early epoch in the evolution of the universe saw the occurrence of a phase transition (or perhaps several) and an exponential growth in size (Starobinsky 1980; Guth 1981). The essential elements of this model are that there exists a scalar field ϕ (or fields) that couple to the other particles, but the importance of this coupling changes with temperature. The basic idea is that the equilibrium value of the field ϕ at fixed temperature $T \neq 0$ is governed by the location of the minimum of the free energy density $V(\phi, T)$, which reduces to the potential of ϕ (i.e., $V[\phi]$) when $T = 0$. A motivation for considering such a scalar field is that in unified theories of the weak, strong, and electromagnetic forces, all vector mesons that mediate these fields are massless and no fundamental differences exist among the interactions when their coupling with the scaler field is not important (or indeed absent). This presumably was the case at the very beginning, when T was so high that the equilibrium value of ϕ was zero. As the temperature dropped, the equilibrium value of ϕ moved away from zero, introducing an important coupling with the other particles that led to some of the vector bosons acquiring mass. Their corresponding interactions became short range, thereby destroying the symmetry between the various forces. During this period, the vacuum energy density presumably did not change as the universe expanded, because of the presence of an ever increasing (nonzero) equilibrium value of ϕ. The effect of this on the expansion of the universe was considerable, leading to an exponential growth in size, i.e., an inflation. Eventually, this energy stored in the field ϕ was transformed into thermal energy, and the universe again became extremely hot, after which its evolution is described by the standard hot universe theory, with the important refinement that the initial conditions for the expansion stage of the hot universe are determined by processes that occurred during the period of inflation. The inflationary model of cosmology still contains some unresolved issues, and variants are now appearing that address some of the remaining problems. Nonetheless, the nearly perfect isotropy of the CMBR shows that the entire observable universe had to be causally connected prior to the time at which the radiation detached from the matter, and this is evidence for an earlier period of inflation.

Still, the CMBR is not perfectly isotropic, even though the distortions are incredibly minute, and this is where much of the current research interest lies. To understand what the full sky maps of the microwave radiation are telling us, let us first consider how variations in its intensity are manifested. The radiation field is generally a function of position and time, and at any given spacetime point has a distribution in both angle and frequency. The specific intensity $I(\mu, \nu)$ is defined to be such that the amount of energy transported by radiation of frequencies $(\nu, \nu + d\nu)$ across a surface element dS, in a time dt, into a solid angle $d\Omega$, is $du = I(\mu, \nu) \, dS \, \mu \, d\Omega \, d\nu \, dt$, where μ is the cosine of the angle between the wave vector (i.e., the direction of travel) and the normal to dS. In cgs units, $I(\mu, \nu)$ has dimensions ergs cm^{-2} s^{-1} Hz^{-1} sr^{-1}. To determine the transformation properties of the specific intensity, we calculate the number of photons N in a frequency interval $d\nu$, passing through an element of area dS oriented perpendicular to the z-axis, into a solid angle $d\Omega$ along an angle $\cos^{-1}\mu$ to the z-axis in a time interval dt. If dS is stationary in the lab frame, we then have

$$N = [I(\mu, \nu)/h\nu] \, (d\Omega \, d\nu) \, (dS \, \mu \, dt) \, . \tag{8.169}$$

An observer in a frame moving with velocity v along the z-axis sees dS moving with a velocity v in the negative z direction. This observer therefore counts

$$N_0 = [I_0(\mu_0, \nu_0)/h\nu_0] \, (d\Omega_0 \, d\nu_0) \, [dS \, \mu_0 \, dt_0 + (v/c) \, dS \, dt_0] \tag{8.170}$$

photons passing through dS. The first term represents the number of photons that would have been counted had dS been stationary, whereas the second term is the photon number density $I_0/ch\nu_0$ times the volume $(dS \, v \, dt_0)$ swept out by dS in a time $dt_0 = \gamma \, dt$. Since both observers must count the same number of photons passing through dS, we must have $N = N_0$, for which

$$I(\mu, \nu) = \left(\frac{\nu}{\nu_0}\right)^3 I_0(\mu_0, \nu_0) \, . \tag{8.171}$$

This follows from the effects of time dilation (see Equation [5.23]) and the transformation of angles (see, e.g., Rybicki and Lightman 1979), which are here conveniently grouped together along with the transformation of frequencies. Clearly, the quantity $I(\mu, \nu)/\nu^3$ is a Lorentz invariant, called the invariant intensity.

The CMBR has a blackbody spectrum described by the Planck law, which in the local co-moving frame is isotropic and may be expressed as

$$I_0(\mu_0, \nu_0) = B_0(\nu_0, T_0) \equiv \frac{2h\nu_0^3/c^2}{\exp(h\nu_0/kT_0) - 1} \, . \tag{8.172}$$

Here, h and k are, respectively, the Planck and Boltzmann constants. We therefore see that according to Equations (5.31) and (8.171), the CMBR intensity measured by an observer at rest in a frame moving with velocity v relative to this co-moving frame must be

$$I(\mu, \nu) = B(\nu, T) \equiv \frac{2h\nu^3/c^2}{\exp(h\nu/kT) - 1} \, , \tag{8.173}$$

where
$$T = \gamma(1 + \beta \mu_0) T_0 . \tag{8.174}$$
In other words, fluctuations in the CMBR that result from local variations in the flow velocity can be characterized as perturbations in the measured temperature of the blackbody spectrum.

After a few months following the Big Bang, the radiation and baryonic matter were strongly coupled by Thomson scattering (see §§ 7.4 and 8.6), which does not alter the photon energy, though it can modify the radiation's angular distribution. A few hundred thousand years later (the exact period depends on cosmological constants that are not known precisely), at a redshift $z \approx 1360$, T fell to the point where helium and hydrogen were about 50% recombined into transparent gases (Peebles 1982). The surface of last scattering occurred a little later, at $z \approx 1160$, after which the photons were free to stream across the universe. Although the rapid scattering between the photons and charged particles prior to recombination effectively isotropized the radiation field, homogeneity could not be maintained by the photons that remained trapped in co-moving coordinates. After the helium and hydrogen recombined to render the medium transparent, any inhomogeneity in the radiation field was manifested as an anisotropy in the CMBR, or equivalently, as a nonuniform temperature T.

There are several effects other than differences in the velocity of the electron-photon fluid that can produce the temperature variations described above. These include fluctuations in (i) the density of the universe at the surface of last scatter, (ii) the gravitational potential of the universe at this surface, and (iii) the gravitational potential along the photon path. Different physical phenomena are responsible for each of these, so that by studying the CMBR, we can infer valuable information about the early universe. For example, in the inflationary cosmology (Starobinsky 1980; Guth 1981), the large-scale structures associated with these anisotropies were once much smaller than the horizon during the inflationary epoch, but grew to be much larger than the horizon as the universe expanded. Indeed, the large-scale inhomogeneities represented by the anisotropy of the CMBR were evidently produced only 10^{-35} seconds after the Big Bang by causal quantum fluctuations.

On smaller scales, inhomogeneities were also produced by "echoes" of the Big Bang. Matter moving at the speed of sound ($c_s \approx c/\sqrt{3}$) had sufficient time before recombination to move the distance spanned by an angle of about 1° on the surface of last scattering. This produced overshooting and the pressure of the photon gas led to oscillations. The temperature variations resulting from this process are also called "acoustic" fluctuations, and the scale associated with how far a sound wave moves from the beginning of the Big Bang to when hydrogen recombines is known as the "sonic" horizon. This distance serves as a ruler for measuring the geometry of the universe. The reason for this is that, while the inhomogeneities associated with primordial density fluctuations are dependent on a number of cosmological parameters, the peak l_{peak} in the Legendre multipole (angular) power spectrum is most sensitive to the total density Ω_0 of the universe. (The expansion of the CMBR distribution in angular space is done by analogy with the multipole expansion for the potential in Equation [8.10], so that l_{peak} corresponds to the dominant spherical harmonic in this series. A fluctuation associated with the lth multipole has a corresponding angular scale of π/l.) This density is expressed as a ratio over the value that characterizes a flat, or "Euclidean," universe. So if $\Omega_0 > 1$, there is sufficient energy in the universe to close it and foretell a future collapse; a value $\Omega_0 < 1$, on the other hand, implies an open universe that should expand forever. At the time of this writing (2000), several observational campaigns have reported the preliminary results of their high-resolution mapping of the CMBR[23] and the results are consistent with inflationary models for the early universe, and a value for Ω_0 of $1.0(+0.15)(-0.30)$. At this point, the data already rule out alternative models, e.g., those based on topological defects, that actively bring about large-scale structure. It should be pointed out, however, that inflationary cold dark matter models also predict the presence of secondary peaks (at higher values of l). The current data limit the amplitude of these harmonics, but they do not exclude their presence, so their existence is still an open question.

Finally, a wealth of information can be found not only in the temperature fluctuations of the microwave background radiation, but also in its polarization characteristics. Scattered light, whether it is sunlight reflected by a haze, or microwave background photons that are reflected off of free electrons in the early universe, is often polarized. This effect occurs because the electron-photon scattering cross section depends upon the polarization of the incoming photon (which corresponds to the direction of the photon's electric field). The physics behind this is similar to that discussed in § 4.3 in the context of a radiation field propagating from one medium to another across an interface, or reflected back by this surface. This effect makes it possible for us to probe the properties of the electrons encountered by the photons as they propagate from the surface of last scatter to our detector.

---

[23]The principal groups that have thus far reported their results are the BOOMERanG experiment (Balloon Observations Of Millimetric Extragalactic Radiation and Geomagnetics), the MAXIMA balloon-borne experiment, and the Mobile Anisotropy Telescope (MAT) at Cerro Toco in Chile. The reader is encouraged to learn more about these discoveries in Miller et al. (1999), de Bernardis et al. (2000), and Hanany et al. (2000).

8.8   COSMIC SUPERLUMINAL SOURCES

Since the 1960s, quasars have been known to emit a significant fraction of their light by synchrotron radiation from relativistic electrons in weak magnetic fields (e.g., van der Laan 1966). In the early days, this model could explain the characteristic radio spectra and the intensity variations on a time scale of years that were observed in some of these sources, but there were problems with this interpretation in others. If the quasar redshifts are indicators of their cosmological distances (which in the 1960s was a subject of some controversy— see Hoyle and Burbidge 1966), then the rapid flickering over several months or less implied that the radiating regions could be no more than a fraction of a light year across. This is based on the simple idea that a significant variation of an object's luminosity can occur only over an interval longer than the light travel time across the source's dimension, for otherwise different portions of the emitter have no way of "communicating" the information that they are changing. Thus, if a fluctuation in the luminosity occurs over a time Δt, it is expected that the source of the variation is no bigger than $\sim c\,\Delta t$. The reason this small size presented a difficulty was that the implied energy density of the particles and the radiation were then much higher than was allowed by the simple synchrotron model. In fact, no radio photons at all would be permitted to escape, since the frequent inverse Compton scatterings with the relativistic electrons would then inevitably lead to the production of high-energy radiation.

An interesting solution to this problem was proposed by Rees (1966, 1967), who suggested that this runaway problem could be alleviated if the rapidly varying sources were expanding with relativistic velocities. Relativistic effects can yield shorter apparent time scales for the variations and permit larger source dimensions. As we shall see below, an additional by-product of this is that parts of the source can then appear to move across the sky with a velocity that exceeds c, i.e., "superluminally." However, the experimental verification of this model could not be obtained until radio astronomy could extend the resolution of the measuring instruments from arcseconds to milliarcseconds. A milliarcsecond corresponds to a length of 1.2 light years at a distance of 100 Megaparsecs, roughly the near edge of the quasar population. This need for higher angular resolution motivated groups in Canada, the United States, England, and the former Soviet Union to develop Very Long Baseline Interferometry (VLBI), in which independent local oscillators are used to obtain and record signals on magnetic tape for later correlation.

During the period between 1968 and 1970, the quasar 3C 279 was monitored by an American-Australian team (Gubbay et al. 1969), who noticed changes in the source visibility and total flux density indicating that during the time of observation, a radio emitting component had reached a diameter of about 1 milliarcsecond, implying expansion at an apparent velocity of at least twice the speed of light. Moffet et al. (1972) concluded that their measurements presented evidence for a relativistic expansion of this component, confirming the basic prediction by Rees several years earlier. The public reaction to the discovery of apparent superluminal motion was generally one of skepticism, inducing some to refute special relativity and/or the concept of cosmological redshifts (e.g., Stubbs 1971). However, in subsequent years, the development of more sophisticated analysis methods for VLBI and the setting up of extensive observing programs have led to the generation of true images of the super-luminal motion of the sources, allowing us to see not only the birth of these components, but also their evolution. Upward of 30 superluminal, or possibly superluminal, quasars have now been cataloged (Zensus and Pearson 1987).

Let us see if we can understand the nature of this phenomenon in a simple way. Suppose we observe a "blob" of relativistic electrons emitting synchrotron radiation (see previous section) as it moves from points 1 to 2 in a time Δt (Figure 8.12). Because 2 is closer to the observer than 1, the apparent duration of the light pulse received at earth is Δt minus the light travel time from point 1 to the line running horizontally out from point 2:

$$\Delta t_{\mathrm{app}} = \Delta t - \frac{\Delta t\, v \cos\theta}{c} = \Delta t \left(1 - \frac{v}{c}\cos\theta\right) . \qquad (8.175)$$

The apparent transverse velocity is therefore

$$v_{\mathrm{app}} = \frac{v\,\Delta t\,\sin\theta}{\Delta t_{\mathrm{app}}}$$

$$= \frac{v\,\sin\theta}{1 - \beta\,\cos\theta} . \qquad (8.176)$$

To find the maximum permissible value, we now differentiate v_{app} with respect to θ and set the result equal to zero. This yields the critical angle θ_c, where

$$\cos\theta_c = \frac{v}{c} \equiv \beta . \qquad (8.177)$$

Correspondingly, $\sin\theta_c = \sqrt{1 - \beta^2} = 1/\gamma$. Therefore, the maximum apparent velocity is

$$v_{\mathrm{max}} = \frac{v\sqrt{1 - \beta^2}}{1 - \beta^2} = \gamma v . \qquad (8.178)$$

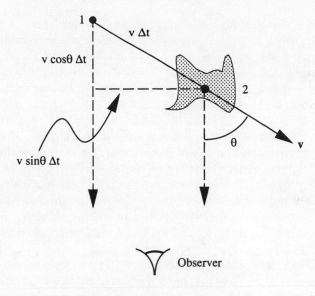

Figure 8.12 A "blob" moving with very high velocity **v** from point 1 to point 2, as seen by a distant observer whose line of sight makes an angle θ with respect to **v**.

Thus, there is a range of angles, all close to zero, for which the *observed* blob velocity greatly exceeds its actual velocity v. Indeed, when $\gamma \gg 1$, v_{max} clearly exceeds c. Does this violate the second postulate of special relativity? Not at all. There is no actual transfer of particles and/or information across the sky. Any two physical points in the blob's path are still connected by an agent traveling with velocity $v < c$.

8.9   POLARIZED RADIATION FROM THE BLACK HOLE AT THE GALACTIC CENTER

The region bounded by the inner few light years at the heart of the Milky Way contains several principal components that coexist within the central deep gravitational potential of the galaxy. Though largely shrouded by the intervening gas and dust, the galactic center is now actively being explored observationally at radio, submillimeter, infrared, X-ray, and γ-ray wavelengths with unprecedented clarity and spectral resolution. The interactions governing the behavior and evolution of this nucleus are attracting many astronomers and astrophysicists interested in learning about the physics of black hole accretion, magnetized gas dynamics, and unusual stellar formation, among others. The galactic center is one of the most interesting regions for scientific investigation because it is the closest available galactic nucleus and therefore can be studied with a resolution that is impossible to achieve in other galaxies.

For example, the galactic center is now known to harbor by far the most evident dark mass condensation, which apparently coincides with a compact, synchrotron-emitting, radio source known as Sgr A*. Several observations (pertaining to the proper and radial motion of stars and gas) strongly support the idea that this object has a mass of over two and a half million suns (Haller et al. 1996; Eckart and Genzel 1996; Ghez et al. 1998). Yet the nature of this unusual object is still largely unknown. One might naively expect from the abundance of nearby gas that this black hole should be accreting prodigiously and be very bright from the conversion of matter energy into radiation. However, it is underluminous (compared to naive expectations) at all wavelengths by many orders of magnitude. Does this imply new physics associated with the accretion of matter through the event horizon, or does it imply something peculiar about the manner with which the gas radiates?

The answers to these questions must be sought from the spectral characteristics of Sgr A*. Of significant current interest is the detection of linear polarization in the millimeter and submillimeter radiation produced by this source. Although the degree of linear polarization in Sgr A* is found to be quite low (less than 1%) below 86 GHz, this is not the case at 750, 850, 1350, and 2000 μm, where a surprisingly large intrinsic polarization of over 10% has now been reported (Aitken et al. 2000). These observations also point to the tantalizing result that the position angle (see § 4.2) changes considerably (by about 80°) between the millimeter and the submillimeter portions of the spectrum, which one would think must surely have something to do with the fact that the emitting gas becomes transparent at submillimeter wavelengths (Melia 1994).

Astrophysicists are now working with the hypothesis that the millimeter and submillimeter radiation from Sgr A* is being produced by hot, magnetized gas orbiting the black hole within a mere 5 to 20 Schwarzschild radii of its event horizon. (A Schwarzschild radius is defined to be $2GM/c^2$, in terms of the black hole mass M and the gravitational constant G. For the black hole at the galactic center, this region is comparable in size to Mercury's orbit.) The reason for this is that although the gas falls in toward the black hole from large distances, it carries some specific angular momentum that forces it to circularize (i.e., to enter into Keplerian rotation) at about 20 Schwarzschild radii from the center (see, e.g., Coker and Melia 1997).

The detection of a net polarization in the emission from Sgr A* is rather strong evidence in favor of this "accretion disk" picture. This is because the circular-ized flow constitutes a magnetic dynamo that greatly amplifies the azimuthal component of the magnetic field. The gas is very hot and highly ionized, so any seed **B** is stretched and wrapped many times by the differentially rotat-ing plasma as it orbits the black hole. The physical state of this gas is such that the dominant radiating mechanism is synchrotron emission (see § 8.6) by a (relativistic) Maxwellian distribution of particles swirling around within this field. Thus, an important constraint arises from the geometry of this highly ordered magnetic field, since it tends to point preferentially in the plane of the disk anchored to the black hole: the polarization vector (i.e., the direction of the electric field associated with the radiation) is generally perpendicular to the azimuthal direction since the synchrotron-emitting particles gyrate around **B**.

Once the radiation is produced, its transport characteristics through the region surrounding the black hole depend sensitively on the wavelength λ. The reason for this is that the photon mean free path is much smaller than the system size above $\lambda = 1$ mm, and larger than the emitting region for $\lambda < 1$ mm. Consequently, the millimeter to centimeter radiation that we see is dominated by emitting elements on the near and far sides of the black hole (since photons from the sides are scattered many times away from the line of sight). For this radiation, the polarization direction tends to be perpendicular to the accretion disk (i.e., to **B** which lies in the plane of the disk). In contrast, the dominant contribution in the submillimeter region comes from the blue-shifted emitter to the side of the black hole, where the Doppler boosting is significant for gas velocities approaching c (see Equation [5.31]). This submillimeter radiation therefore has a polarization vector pointing along the disk as seen at earth.

The net result of these effects is that the radio emission from Sgr A* is partially polarized, with a position angle that changes by almost 90° from the millimeter to submillimeter portions of the spectrum. Polarization measurements such as these may eventually constitute a powerful diagnostic of the emitting region just outside the event horizon of the black hole at the galactic center and provide us with an opportunity of studying in detail the predictions of general relativity in the strong field limit.

REFERENCES

Aitken, D. K., Greaves, J., Chrysostomou, A., Jenness, T., Holland, W., Hough, J. H., Pierce-Price, D., and Richer, J., *Detection of Polarized Millimeter and Submillimeter Emission from Sgr A**, Astrop. Journal Letters 534 (2000), pp. L173-L176.

Alvarez, L. W., Antuna, Jr., M., Byrns, R. A., et al., *A Magnetic Monopole Detector Utilizing Superconducting Elements,* Rev. Sci. Instr. 42 (1971), pp. 326-330.

Barut, A. O., *Electrodynamics and Classical Theory of Fields and Particles,* Macmillan, 1964.

Blatt, J. M., and Weisskopf, V. F., *Theoretical Nuclear Physics,* Wiley, 1952.

Borgnis, F. E., and Papas, C. H., *Electromagnetic Waveguides and Resonators,* Vol. XVI *Encyclopaedia of Physics,* Springer, 1968.

Born, M., and Wolf, E., *Principles of Optics,* 4th edition, Pergamon, 1970.

Bouwkamp, C. J., and Casimir, H. B. G., *On Multipole Expansions in the Theory of Electromagnetic Radiation,* Physica 20 (1954), pp. 539-554.

Butkov, E., *Mathematical Physics,* Addison-Wesley, 1968.

Byckling, E., and Kajantie, K., *Particle Kinematics,* Wiley, 1973.

Cabrera, B., *First Results from a Superconductive Detector,* Phys. Rev. Letters 48 (1982), pp. 1378-1381.

Churchill, R. V., *Fourier Series and Boundary Value Problems,* 2nd edition, McGraw-Hill, 1963.

Coker, R. F., and Melia, F., *Hydrodynamical Accretion onto Sgr A* from Distributed Point Sources,* Astrop. Journal Letters 488 (1997), pp. L149-L153.

Courant, R., and Hilbert, D., *Methods of Mathematical Physics*, Wiley-Interscience, 1962.

Davydov, A. S., *Quantum Mechanics*, Pergamon Press, 1976.

de Bernardis, P., et al., *A Flat Universe from High-Resolution Maps of the Cosmic Microwave Background Radiation*, Nature 404 (2000), pp. 955-959.

Eckart, A., and Genzel, R., *Observations of Stellar Proper Motions near the Galactic Centre*, Nature 383 (1996), pp. 415-417.

Fixsen, D. J., Cheng, E. S., Gales, J. M., Mather, J. C., Shafer, R. A., and Wright, E. L., *The Cosmic Microwave Background Spectrum from the Full COBE/FIRAS Data Set*, Astrop. Journal Letters 473 (1996), pp. L576-L579.

Fox, J. G., *Evidence Against Emission Theories*, American J. Phys. 33 (1965), pp. 1-17.

Fox, J. G., *Constancy of the Velocity of Light*, J. Opt. Soc. 57 (1967), pp. 967-968.

Ghez, A. M., Klein, B. L., Morris, M., and Becklin, E. E., *High Proper-Motion Stars in the Vicinity of Sgr A*: Evidence for a Supermassive Black Hole at the Center of Our Galaxy*, Astrop. Journal 509 (1998), pp. 678-686.

Goldhaber, A. S., and Trower, W. P., *Resource Letter MM-1; Magnetic Monopoles*, Am. J. Phys. 58 (1990), pp. 429-439.

Goldstein, H., *Classical Mechanics*, Addison-Wesley, 1980.

Groom, D., *In Search of the Supermassive Magnetic Monopole*, Phys. Reports 140 (1985), pp. 323-373.

Gubbay, J., Legg, A. J., Robertson, D. S., Moffet, A. T., Ekers, R. D., and Seidel, B., *Variations of Small Quasar Components at 2,3000 Hz*, Nature 224 (1969), pp. 1094-1095.

Guth, A., *Inflationary Universe: A Possible Solution to the Horizon and Flatness Problems*, Phys. Rev. D 23 (1981), pp. 347-358.

Haller, J. W., Rieke, M. J., Rieke, G. H., Tamblyn, P., Close, L., and Melia, F., *Stellar Kinematics and the Black Hole in the Galactic Center*, Astrop. Journal 456 (1996), pp. 194-205.

Hanany, S., et al., *MAXIMA-1: A Measurement of the Cosmic Microwave Background Anisotropy on Small Angular Scales*, Astrop. Journal Letters 545 (2000), pp. L5-L8.

Hay, H. J., Schiffer, J. P., Cranshaw, T. E., and Egelstaff, P. A., *Measurement in the Red Shift in an Accelerated System Using the Mössbauer Effect in* Fe^{57}, Phys. Rev. Letters 4 (1960), p. 165.

Heitler, W., *Quantum Theory of Radiation,* 3rd edition, Oxford University Press, 1954.

Hoyle, F., and Burbidge, G. R., *On the Nature of the Quasi-Stellar Objects,* Astrop. Journal 144 (1966), pp. 534-552.

Iwanenko, D., and Sokolow, A., *Klassische Feldtheorie,* Akademie-Verlag, 1953.

Jackson, J. D., *Classical Electrodynamics,* 3rd edition, John Wiley and Sons, 2000.

Jaseva, T. S., Javan, A., Murray, J., and Townes, C. H., *Test of Special Relativity or of the Isotropy of Space by the Use of Infrared Masers,* Phys. Rev. 133A (1964), pp. 1221-1225.

Lamb, W. E., and Retherford, R. C., *Fine Structure of the Hydrogen Atom by a Microwave Method,* Phys. Rev. 72 (1947), pp. 241-243.

Landau, L. D., and Lifshitz, E. M., *The Classical Theory of Fields,* 4th edition, Pergamon Press, 1975a.

_____ , *Electrodynamics of Continuous Media,* 3d edition, Pergamon Press, 1975b.

Linde, A., *Particle Physics and Inflationary Cosmology,* Harwood Academic Publishers, 1990.

Lorentz, H. A., *Theory of Electrons,* 2nd edition, Dover, 1952.

Magnus, W., Oberhettinger, F., and Soni, R. P., *Formulas and Theorems for the Special Functions of Mathematical Physics,* Springer-Verlag, 1966.

Maxwell, J. C., *Treatise on Electricity and Magnetism,* 3rd. edition, Dover, 1954.

Melia, F., *An Accreting Black Hole Model for Sgr A*. 2: A Detailed Study,* Astrop. Journal 426 (1994), pp. 577-585.

Michelson, A. A., and Morley, E. W. *On the Relative Motion of the Earth and the Luminiferous Ether,* American J. Science 34 (1887), pp. 333-341; reprinted in *Relativity Theory: Its Origins and Impact on Modern Thought,* ed. by L. P. Williams, John Wiley and Sons, 1968.

Miller, A. D., et al., *A Measurement of the Angular Power Spectrum of the CMB from l = 100 to 400,* Astrop. Journal Letters 524 (1999), pp. L1-L4.

Moffet, A. T., Gubbay, J., Robertson, D. S., and Legg, A. J., in *IAU Symposium 44, External Galaxies and Quasi-Stellar Objects,* ed. D. S. Evans, Reidel, 1972.

Morse, P. M., and Feshback, H., *Methods of Theoretical Physics,* 2 Pts., McGraw-Hill, 1953.

Panofsky, W. K. H., and Phillips, M., *Classical Electricity and Magnetism,* Addison-Wesley, 1962.

Parker, E. N., *Cosmical Magnetic Fields: Their Origin and Their Activity,* Oxford University Press, 1979.

Peebles, P. J. E., *Large-Scale Background Temperature and Mass Fluctuations Due to Scale-Invariant Primeval Perturbations,* Astrop. Journal Letters 263 (1982), pp. L1-L4.

Penzias, A. A., and Wilson, R. W., *A Measurement of Excess Antenna Temperature at 4080 Mc/s,* Astrop. Journal 142 (1965), pp. 419-421.

Ramo, S., Whinnery, J. R., and Van Duzer, T., *Fields and Waves in Communication Electronics,* Wiley, 1965.

Rees, M. J., *Appearance of Relativistically Expanding Radio Sources,* Nature 211 (1966), pp. 468-470.

_____ , *Studies in Radio Source Structure,* Monthly Notices Roy. Astron. Soc. 135 (1967), pp. 345-360.

Rohrlich, F., *Classical Charged Particles,* Addison-Wesley, 1965.

Rybicki, G. B., and Lightman, A. P., *Radiative Processes in Astrophysics,* Wiley-Interscience, 1979.

Sakurai, J. J., *Advanced Quantum Mechanics,* Addison-Wesley, 1973.

Sard, R. D., *Relativistic Mechanics,* Benjamin, 1970.

Schott, G. A., *Electromagnetic Radiation and the Mechanical Reactions Arising from It,* Cambridge University Press, 1912.

Schwinger, J., *A Magnetic Model of Matter,* Science 165 (1969), pp. 757-761.

Smythe, W. R., *Static and Dynamic Electricity,* 3rd edition, McGraw-Hill, 1969.

Sommerfeld, A., *Partial Differential Equations,* Academic Press, 1949.

Starobinsky, A. A., *A New Type of Isotropic Cosmological Models without Singularity,* Phys. Letters B 91B (1980), pp. 99-104.

Stubbs, P., *Red Shift Without Reason,* New Scientist 50 (1971), pp. 254-255.

Van Bladel, J., *IEEE Antennas and Propagation Magazine,* 33, no. 2 (April 1991), pp. 69-74.

van der Laan, H., *A Model for Variable Extragalactic Radio Sources,* Nature 211 (1966), pp. 1131-1133.

Weinberg, S., *Gravitation and Cosmology: Principles and Applications of the General Theory of Relativity,* Wiley, 1972.

Williams, E. J., *Correlation of Certain Collision Problems with Radiation Theory,* Kgl. Danske Videnskab. Selskab Mat.-fys. Medd. XIII, no. 4, 1935.

Williams, L. P., *The Origins of Field Theory,* Random House, 1966.

Yang, C. N., and Mills, R. L., *Conservation of Isotopic Spin and Isotopic Gauge Invariance,* Phys. Rev. 96 (1954), pp. 191-195.

Yndurain, F. J., *Quantum Chromodynamics: An Introduction to the Theory of Quarks and Gluons,* Springer-Verlag, 1983.

Zensus, J. A., and Pearson, T. J., *Superluminal Radio Sources,* Cambridge University Press, 1987.

INDEX